Information Technology & Library Services

Information Technology & Library Services

Editor

Silvius Mateus

scitus
academics

Information Technology & Library Services

Edited by **Silvius Mateus**

ISBN: 978-1-68117-558-4
Library of Congress Control Number: 2016942967

SCITUS Academics LLC,
www.scitusacademics.com
Box No. 4766, 616 Corporate Way,
Suite 2, Valley Cottage,
NY 10989

Printed in United States of America on Acid Free Paper

Preface

The organization of information/knowledge is an essential preliminary to its effective exploitation and dissemination. As the quantity of knowledge expands, the need to organize it becomes more pressing. A vast number of different means of organizing information have been devised and exploited since the earliest times. Information technology (IT) is the modern buzz word, it has provided facilities for the free flow of information. The world has become a global village with information superhighways created through networks like Internet. This has facilitated electronic librarian-ship with the diversification of library facilities and services to its user. The concept of virtual libraries OPAC, Hypertext, and teleconferences for the purpose of library and information services have become common. New technological developments have already profoundly affected libraries; almost every function carried out in a library has been altered to some extent by advances in electronics, computerization, and telecommunications. Information Technology as the coming together of computing and telecommunications for the purpose of handling information; the application of technologies to information handling; including generation, storage, processing, retrieval and dissemination. It is also concerned with the acquisition, processing, storage and dissemination of information-textual, numerical, pictorial and vocal. It is a broad-based term comprising the gathering, organization (packaging), storage and retrieval (dissemination) of information that can be in textual or numerical (books, documents), pictorial and vocal forms (audio–visual) or a combination of all the above (multimedia), using a combination of computer and telecommunications devices.

Information Technology & Library Services include libraries and the internet, the concept of the digital library, impact on professional roles, communicating, changing perspectives on service provision, instruction and training, and networking, data communication. This book identifies emerging technologies and assesses their impact on reference service, bibliographic instruction, and organizational management.

It becomes explicit from the above that IT in libraries comprises all the electronic infrastructure and facilities employed by libraries to improve and provide efficient services. Such facilities, in broad term, consist of hardware, software and communication links between the service outlets of different libraries to facilitate the sharing of common resources; especially the library networks.

Table of Contents

Chapter 1

Costs and Benefits of a Shared Digital Long-Term Preservation System

Esa-Pekka Keskitalo

National Library of Finland, PL 26, 00014 University of Helsink

ABSTRACT

This paper describes the cost-benefit analysis of digital long-term preservation (LTP) that was carried out in the context of the Finnish National Digital Library Project (NDL) in 2010.

The analysis was based on the assumption that as many as 200 archives, libraries, and museums will share an LTP system. The term 'system' shall be understood as encompassing not only information technology, but also human resources, organizational structures, policies and funding mechanisms.

The cost analysis shows that an LTP system will incur, over the first 12 years, cumulative costs of €42 million, i.e. an average of €3.5 million per annum. Human resources and investments in information technology are the major cost factors. After the initial stages, the analysis predicts annual costs of circa €4 million.

The analysis compared scenarios with and without a shared LTP system. The results indicate that a shared system will have remarkable benefits. At the development and implementation stages, a shared system shows an advantage of €30 million against the alternative scenario consisting of five independent LTP solutions. During the later stages, the advantage is estimated at €10 million per annum. The cumulative cost benefit over the first 12 years would amount to circa €100 million.

1. INTRODUCTION

Securing LTP requires planning of sustainable models for costs and funding that ensure the usability of information. The way of action must balance costs and benefits over a long period of time.

Digital preservation is characterized by its active nature. According to present understanding, digital materials cannot survive without constant assessment and prevention of risks. Resources, from electricity to highly skilled IT specialists, will be needed permanently.

A better understanding of the funding challenges of the future is urgently needed in order to make informed decisions about the direction of digital preservation — and in order to explain the feasibility, if any, of digital preservation to the decision-makers.

2. BACKGROUND

2.1 The National Digital Library Project

The NDL Project was launched in the Summer of 2008 by the Ministry of Education and Culture. Its objective is to improve availability and usability of digital materials offered by archives, libraries, and museums; and to develop an LTP solution for these materials. The Project is part of broader efforts to improve the national digital infrastructure and services.

The ultimate target of the NDL is to provide better conditions for research, learning, and general information retrieval, as well as to promote art and creativity. In addition, it pursues better productivity of the participating organizations. It is believed that through shared solutions it is possible to reduce overlapping costs in digitization, management of digital materials, and creating digital services, thus enabling reassignment of resources and more rational use of public funds. The NDL project is also aligned with the European Union's objectives for digitization of cultural materials and scientific information, their digital availability, and their long-term preservation.

The project has two major branches, or sections: the section of 'the public interface', and the section of 'long term preservation' (see Figure 1). This paper focuses solely on the latter [The entire NDL project was presented to the LIBER Annual Conference in 2010 (Hormia, 2010)]. The enterprise architecture of the NDL was published in 2010 (Enterprise, 2010).

2.2 Long-Term Preservation in the NDL Project

The idea of extensive cooperation between the archive, library and museum sectors on LTP goes back several years. In February 2007, the Ministry of Education and Culture set up a working group with the task to draw up national principles of LTP, and to prepare a first draft for organizing LTP, paying special attention to the benefits of a shared infrastructure and centralized services. The working group reported in January 2008. It found that LTP is a challenge that must be looked at as a whole and that shared LTP systems and services are both feasible and desirable.

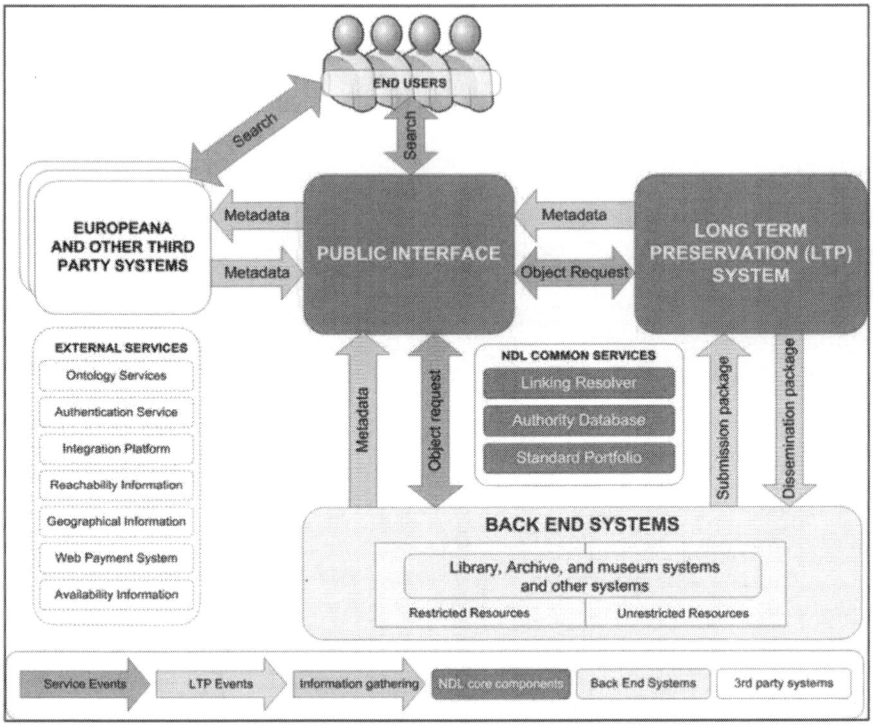

Figure 1. The National Digital Library.

The LTP Section of the NDL Project continued along the lines of the 2007 working group. A report was submitted in June 2010 (Long-Term, 2010). The cost-benefit analysis discussed here is an annex to that report. The work on LTP has continued after the report with specialists preparing more detailed plans for the LTP system, its administration, etc. Final decisions on funding were still pending when this article was submitted.

The cost-benefit analysis was carried out by CSC — IT Centre for Science Ltd (www.csc.fi). It is a state-owned company that provides IT support and resources for universities, research institutes and companies. The analysis was supported by a number of surveys and sub-projects carried out by the NDL Project.

The author of this paper was closely involved in the process, being the representative of the National Library of Finland in the LTP Section of the NDL Project organization, and acting as a liaison with the library sector in preparing the cost-benefit analysis and supporting surveys.

2.3 Cost Analysis and Benefits Analysis

Economic theory offers many tools for cost accounting. The foundation of cost accounting is an understanding, as thorough as possible, of all cost factors and cost effects of a function through its life cycle.

The analysis of life cycle costs is an attempt to understand thoroughly the total life cycle of an operation and use the information in comparing options. Life cycle analysis also assists in timing and prioritizing future investments.

Cost-benefit analysis is a tool of financial decision-making. It can be used to compare options and to recognize beneficial and harmful factors in the operational environment. For example in policymaking, economic and environmental benefits are converted into money values for easier comparison.

The cost analysis conducted in the NDL Project exploited the LIFE model (Lifecycle Information for E-Literature model) (Life, 2008) with modifications. The LIFE model has been created for the purpose of estimating storage costs of digital collections.

The Standard Cost Model for Assessment of Citizens' Benefits was used in analysing the benefits. It has been used in other evaluations of governmental IT projects in Finland, too. The model identifies qualitative and cost benefits not only for the organizations but also for the citizens, taking into account the number of transactions per citizen per annum, and the time and other resources consumed by a transaction. In this model, the emphasis of the benefit analysis was put on monetary benefits. The model identifies benefit factors and estimates their importance as well as the speed and probability of their realization. [For an introduction in English, see e.g. (van den Hurk, 2008).]

2.4 International Comparisons

LTP is a subject of avid interest and a number of projects around the world. National and other large-scale approaches are not uncommon. However, as far as costs and cost-benefit analyses were concerned, specific information was hard to come by: data were often scattered, preliminary, or they could not be disclosed. The case studies of the LIFE project were important, as was information obtained from the National Digital Heritage (NDHA) programme of New Zealand.[1]

2.5 Basic Assumptions

2.5.1 The Concept of a Shared LTP System

The aim of the NDL project is to create a centralized system for digital LTP. This system should be able to provide preservation services to different kinds of organizations, curating different types of materials and having different customers.

The 2007 working group studied several options quite thoroughly and ended up recommending a centralized model. This model has also earned wide support among organizations participating in the NDL Project.

The cost-benefit analysis used a hypothetical model of two geographically separate places of operation that both have facilities for ingest, preservation, and dissemination of digital materials.

In the analysis, LTP is always understood to cover three aspects of preservation, namely

- bit-level preservation,
- preservation of interpretability, and
- preservation of original experience.

More information about the concepts used in the analysis is available in the Final Report of the LTP Section (Long-Term, 2010).

2.5.2 Users of the LTP System

The NDL is intended for archives, libraries, and museums supervised by the Ministry of Education and Culture. There are about two hundred such organizations. On the national level they include the Finnish National Archives Service, the National Library of Finland, the Research Institute for the Languages of Finland, the National Board of Antiquities, the Finnish National Gallery, the National Audiovisual Archive, and the Finnish Museum of Natural History.

2.5.3 Amount of Data

In 2009, two surveys were carried out to take stock of the size of the digital collections that might potentially be ingested into the LTP system.

The estimated amount of data for the present and near future is shown in Table 1. Typically, a small number of organizations provide most of the material in the library and archive sectors respectively, whereas in the museum sector digital materials are more evenly distributed over the organizations.

Table 1. Materials in the NDL, estimated in 2008

Type	Amount of data (Terabytes)			
	2008	2009	2010	2011
Text documents	64	164	190	216
Still images	6	10	17	25
Moving image	11	25	31	37
Audio	17	23	28	33
Reference records	<10	<10	<10	<10
Web Archive	8	17	26	35
Radio and TV Archive	2	57	112	167
Total	108	296	404	513

A question about the expected growth of digital collections from 2008 to 2025 yielded the following, very tentative, figures:

- archives, from 250 Tb to 3,000 Tb;
- libraries, from 65 Tb to 600 Tb; and
- museums, from 14 Tb to 270 Tb

All respondents felt very unsure about these figures. It must also be noted that data from scientific research are under-represented in the figures. It appears that at least the amount of scientific data in need of preservation is underestimated. Also, the survey omitted all potential partners not supervised by the Ministry of Education and Culture. Their inclusion could change the numbers significantly.

Parts of the surveys were repeated in late Spring 2011. According to the preliminary analysis some organizations have raised their estimates substantially. The study also showed that services related to emulation as a preservation strategy are not expected from the shared LTP system by potential participating organizations. The results will be published later in 2011.

2.5.4 Construction Stage of an LTP System

For the purposes of the cost analysis, it was necessary to make assumptions as to how an LTP system will take shape. We assumed that the system would take four years to go into production, and that it would be completed by year seven. Year-by-year assumptions as to the number of participating organizations and the amount of data were also necessary. These are summarized in Table 2.

Table 2 Assumed Building Phases of the shared LTP System

Year	Phase	Organizations	Materials
1	Functional requirements completed; developing tools and supporting services; planning acquisitions		
2	Developing tools and supporting services, putting out tenders, start piloting		
3	First back-end system connected; integration project; going into production	5 organizations 5 systems	300 Tb
4	1st production phase: first preservation location in production. — Increasing capacity; use and maintenance.	20 organizations 10 systems	700 Tb
5	1st production phase. — Preparing geographical expansion. 2nd phase tenders.	80 organizations 20 systems	1,000 Tb
6	2nd production phase: two locations in production. — Increasing capacity.	140 organization 30 systems	1,400 Tb
7–12	2nd production phase. — Capacity increases 15–25% annually; updating hardware and software; at year 11, replacing LTP software.	209 organizations 40 systems	4,000 Tb

2.5.4 Costs of the Shared LTP System: Methods

The LIFE model was the starting point for analyzing the cost of the shared LTP System. It is a life-cycle model for assessing present and future costs of LTP. The model identifies six stages of the life cycle —

Creation/Purchase, Acquisition, Ingest, Bit-Stream Preservation, Content Preservation, and Access. These stages help to position costs on a time scale and identify cost peaks. The LIFE model required slight modifications for the purposes of the present analysis. The modifications mainly concerned the question of repeated stages: in the transfer of materials to the shared LTP systems some components of the Ingest stage are repeated.[2]

2.5.5 Costs of Digital Preservation

General Assumptions Concerning Costs

The starting point of the analysis was that a shared LTP system would be built. The general architecture of the system was fairly well planned and agreed upon by the time the analysis was done.

Cycles of Hardware and Software Replacement

- Increase of disk space, 1 year
- Disk arrays, 3 years
- Servers, 3–5 years
- Tape robot, 5–8 years
- Tape drives, 3–5 years
- Network, 3–5 years
- System administration and control software, 3–8 years
- LTP software, 5–10 years

- Format-dependent software (for accessing preserved materials), 3–5 years

Amount of Materials to be Preserved

Building on the basic assumptions (above) it was further postulated that:

- All materials will be transferred to the LTP, although gradually.
- By end of 2011 the amount of materials is 700 Terabytes.
- The amount of materials will increase by 15% annually.
- File format migrations will happen every ten years for all kinds of materials, i.e. 10% of the total collection will be migrated every year. Both versions are supposed to be of equal file size, and both will be preserved. Ten percent more storage space will be needed every year because of migrations.
- Updates (e.g. revisions of metadata) usually do not necessitate copying the objects and thus do not increase the need for storage space.

Cost of Human Resources

General statistics of 2009 were used as the source for HR costs and their rise over time. Further estimations were made about the costs of the workspace, employers' additional costs, costs of outsourced services, etc. The exact figures are very much dependent on local conditions, so they are not presented here.

Number of Organizations Using the Shared LTP System

It was assumed that all potential libraries, archives and museums would join the shared LTP System. Therefore, the following estimates were made of the number of participating organizations:

- Libraries: 30 organizations, 10 systems
- Archives: 15 organizations, 15 systems
- Museums: 164 organizations, 15 systems
- Total: 209 organizations, 40 systems.

Costs of the Ingest Stage

Work done in the participating organizations was excluded from the calculations. The role of the LTP System is considered to be in support and consultation. After the initial phase, one person-year was thought to be sufficient.

Costs of the Bit-Level Preservation Stage

Materials should be stored in several copies, on different media and in different locations. In the hypothetical model, there were (from year six on) full LTP services available at two separate locations, each keeping digital objects at least in three copies on at least two types of media. The costs of one additional dark archive were calculated, too. The dark archive was understood to be a geographically separate place of storage only to be used internally by the LTP system.

The access time to any materials was allowed to be ‘a few seconds’ at most.

Capacity, Purchase Costs, and Maintenance Costs of Storage Media

These costs were predicted to go down:

- The price of storage tape (relative to capacity) will drop by half every three years.
- The price of hard disks (relative to capacity) will drop by half every two years.
- However, same amount of money will be invested in new models of storage systems, thus increasing the capacity.
- From year seven on, the annual costs of the storage systems will remain stable. The demand for more space will be covered by decreasing prices.
- Consumption of electricity will decrease by 10% per byte per annum.

2.6 Other Assumptions

Maintenance of hardware, operating systems etc. will take five person-years, the dark archive adding one person-year.

There will be an 'LTP application' that provides tools for integrity control, digital signatures management and preservation procedures (such as file format migrations). Various ways of obtaining such a software system (licencing, building one internally, or any combination of these) were assumed to cost the same.

Costs of the Content Preservation Stage

It is important to note that an exact division of labour greatly influences the costs, as those costs incurred by the participating organizations are not shown.

- It is assumed that materials will be ingested in the shared LTP system in an up-to-date format. File format migrations will become necessary by year eleven.

- Ten percent of the materials will be migrated to a new file format every year.

- There are 20 major types of file formats. Assessing their status and migrations will take about six person – years from year eleven on.

- Manual checking and assessing of materials is understood to be carried out by the participating organizations.

Costs of the Access Stage

These are thought to be relatively low.

The annual total costs of the shared LTP system are shown in Figure 2. The main breakdown of the costs is shown in Figure 3.

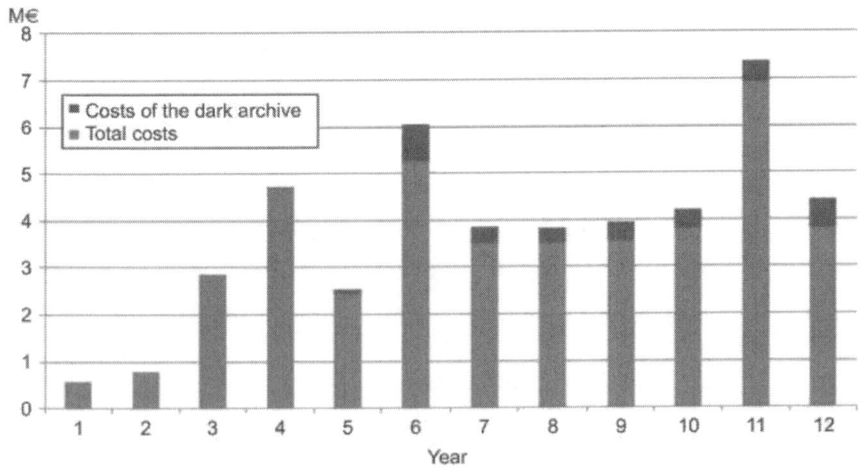

Figure 2. Total Costs of the Shared LTP System.

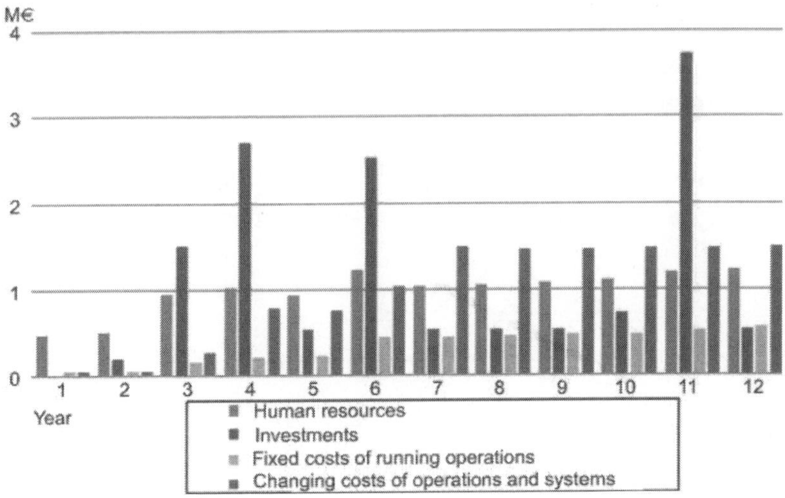

Figure 3. Main Breakdown of Costs of the Shared LTP System.

3. QUALITATIVE BENEFITS OF DIGITAL PRESERVATION AND THE SHARED LTP SYSTEM

In choosing the methods, the NDL Project was informed by the eServices and eDemocracy Acceleration Programme (SADE), run by the Ministry of Finance. The SADE Programme has created a procedure for assessment of benefits, based among others on the Standard Cost Model for Citizens. According to the SADE procedure, the project focussed was on benefits the costs and benefits of which could be measured in money.

In the SADE procedure, a benefit index is calculated. On an a scale from 0 to 100, it reflects

- the importance of a beneficial factor,
- the time span of the realization of the benefit, and
- the probability of realization.

None of the benefits of a shared LTP system achieved the highest scores. It must be noted that the model is designed to strongly favour fast implementation and quickly realized benefits.

The analysis of the concept of a shared LTP system identified eight qualitative benefits, listed in Table 3 with their benefit index numbers.

Table 3. Qualitative Benefits of a Shared LTP System

Benefit	Importance	Time scale	Probability	**Benefit index**
(1) It supports organizations in fulfilling their legal obligations.	3	2	3	**81**
(2) It helps organizations to focus on core functions.	3	2	3	**81**
(3) It reduces overlapping operations.	2	2	3	**63**
(4) It makes it easier to bring skills and knowledge together so that they are not dependent on one person.	2	2	3	**63**
(5) It reduces risks of failure in digital preservation	3	1	3	**63**
(6) It creates better processes and services.	2	2	3	**63**
(7) It makes it possible for future generations to use and reuse the materials preserved.	3	1	3	**63**
(8) It enables seamless cooperation and share of resources across organizational borders.	3	1	2	**38**

4. QUANTITATIVE BENEFITS

Central to the SADE model is the translation of qualitative benefits into amounts of money. These monetary benefits may be divided into

- savings in performance;
- savings in comparison to alternative modes of operation;
- increased profits; and

- other savings.

In the case of the shared LTP system, the most visible monetary benefits are the following:

Savings in Using Shared Solutions (benefits 3 and 4)

The analysis indicates that savings of circa €30 million during the planning and implementation stages might be obtained by adopting the shared LTP system. When in production, the savings would be around €8 million per annum.

These calculations assume an alternative situation with five separate LTP systems. Savings are obtained in both human resources and in investments. For example, as software costs do not increase with the amount of materials preserved in a system, they are multiplied in the alternative scenario.

Savings in Costs of Preservation Management (benefits 2 and 6)

Savings in performance during the production phase are estimated at €2.5 million per annum. It is assumed that in the alternative scenario every organization designs processes for management, distribution and quality assurance, and that every organization allocates to these functions 60% of the hours of 2 persons.

Savings Created by the Value of the Digital Materials Being Preserved (benefit 5)

The savings created by the fact that digital preservation is indeed taken care of are shown to be €0.5 million per annum. Digital preservation produces benefits through reuse of materials in education, creative activities and research, as well as in digital products and services based on the materials preserved.

For purposes of the analysis, it was estimated that two percent of digitizedmaterials would be destroyed or damaged annually without an LTP solution and thus should have to be redigitized. The cost of digitization was put at €1.3 per object. [The figure was based on data in (Numeric, 2009), and information gathered from the organizations participating in the NDL Project.] The benefits of digitization per se (rather than preservation of the results), such as reduced costs of stacks and easier access, were not taken into consideration.

As far as born-digital materials were concerned, the savings were believed to be 'several million' euros per annum. As above, the rate of deterioration was assumed to be two percent annually. For many born-digital materials it is impossible to recreate them once they are damaged. The savings are the result of a lessening burden of curating damaged materials, and of the productive use of materials preserved. Monetary losses caused by damage to born-digital materials were, of course, hard to calculate. It was postulated that the costs are significantly higher than those of redigitization.

The final analysis shows a benefit of €30 million in the development and implementation phases when using a shared LTP system, compared to a model with many independent systems. During the production, the savings are estimated at €10 million annually.

5. CONCLUSIONS

Cost-benefit analysis of LTP proved difficult. That hardly surprised anyone, for the time-span of analysis is very long; future developments in information technology are relatively hard to predict; and comparable data from other projects were scarce. Lastly, the organizations that provided information for the analysis felt very insecure about their prognoses.

Throughout the analysis, it was obvious that mainstream tools of cost-benefit analysis are not particularly suited to an analysis of LTP, as they showed a tendency to favour quick implementation and quick returns on investment.

Nevertheless, even allowing for a wide margin of error, the results of the analysis are compelling. They certainly seem to confirm the widely held conviction that extensive cooperation of archives, libraries, and museums makes sense — at least in the field of LTP and at least from a financial point of view.

The main lesson from the NDL Project was that cost and benefit analysis of LTP could, in fact, be done. The particular figures from this exercise very probably become obsolete in the near future — some are obsolete already. A more important and more permanent result is the framework for thinking about costs and benefits.

The next round will be easier.

REFERENCES

1. Enterprise, 2010: The National Digital Library project/Ministry of Education andCulture: The National Digital Library — Enterprise Architecture. V1.0. 2010. http://www.kdk.fi/en/ enterprise-architecture.

2. Hormia, 2010: Hormia, Kristiina: Libraries, archives and museums working together! Learning by doing! Making the collections and services of libraries, archives and museums digitally available. Presentation slides, LIBER 2010. http://www.statsbiblioteket. dk/ liber2010/presentations/Hormia.pdf.

3. Life, 2008: Ayris, P., Davies, R., McLeod, R., Miao, R., Shenton, H., Wheatley, P. The LIFE2 final project report. The LIFE2 Project. 2008. http://www.life.ac.uk/2/ documentation.shtml.

4. Long-Term, 2010: The National Digital Library project/Ministry of Educationand Culture: The National Digital Library Initiative — Long-Term Preservation Project. Final Report. V. 1.0. 2010. http://www.kdk.fi/images/stories/LTP_Final_Report_v_1_1.pdf.

5. Numeric, 2009: Chartered Institute of Public Finance and Accountancy (CIPFA): NUMERIC — Developing a statistical framework for measuring the progress made in the digitisation of cultural materials and content. Study deliverable No 8: Study Report. Study findings and proposals for sustaining the framework. May 2009. http://cordis.europa.eu/fp7/ict/telearn-digicult/numeric-study_en.pdf.

6. van den Hurk, 2008: Joey van den Hurk, Peter Rem, Milan Jansen: Standard Cost Model for Citizens. User's guide for measuring administrative burdens for Citizens. Ministry of the Interior and Kingdom Relations, The Hague 2008. http://www. whatarelief.eu/publications/standard-cost-model.

NOTES

[1] http://ndha-wiki.natlib.govt.nz/ndha/

[2] Unlike in the original model, in the analysis described here the Acquisition stage is understood to cover all steps before the transfer, including cataloguing. Ingest, on the other hand, only refers to the ingest into the LTP system.

Chapter 2

Designing Libraries for Research Collaboration in the Network World: An Exploratory Study

Sheila Corrall

University of Pittsburgh, USA

ABSTRACT

Advances in technology and changes in higher education and research are forcing libraries to rethink the services they offer to researchers and how they are presented. Librarians are moving into "higher end support" and adopting new service models based on "deep collaboration" with academic partners. The internal arrangements that form the design of an organization can determine whether it achieves its mission under particular conditions. The present study explores how libraries in 24 leading UK research universities are organizing resources and services to support the research enterprise. Qualitative data were collected from institutional websites and other public domain sources and were analysed using matrix techniques. The results show a trend away from integrated library and computing service organizations;

variety in institutional reporting lines, but predominantly in large professional service groups; consistency in internal library groupings, but variation in portfolio and job titles; expansion of specialist positions, with new functional roles complementing traditional subject liaisons; and dedicated spaces, working groups, and integrated websites promoting boundary-spanning activities. The findings confirm and extend prior work and are being used to design a large scale international survey.

KEYWORDS

management structures, organization design, research support, university libraries

1. INTRODUCTION

Developments in digital media, network technologies, research workflows, scholarly communication and funding policies are challenging academic libraries to respond with a wider array of services and facilities for researchers (**Bourg, Coleman, & Erway, 2009**; **Housewright, Schonfeld, & Wulfson, 2013a**; **Vinopal & McCormick, 2013**). Subject/liaison librarians have traditionally provided research services as part of their academic support role, but libraries are increasingly identifying research support as a specific area requiring additional co-ordination and strategic development:

*"Open-access publishing, institutional repositories, the need to co-ordinate collection management and storage, the increasing availability of information technology (IT) tools to help researchers with their work and, not least, the Research Excellence Framework (REF) have all pointed to the need for a coherent library strategy and response" (**Bradbury & Weightman, 2010, p. 65**).*

University library support for research has emerged as a strategic issue and generated studies investigating service developments, new roles, skills requirements (**Auckland, 2012**; **Corrall, Kennan, & Afzal, 2013**; **Covert-Vail & Collard, 2012**; **Jaguszewski & Williams, 2013**). The present paper offers a different perspective on the topic by examining library engagement with research and researchers through the lens of organizational design.

1.1 Background

There is a perceived requirement for libraries to develop more specialized interventions to provide point-of-need research support and advice around the whole knowledge creation cycle: from ideas generation and project conception, through data acquisition, manipulation and interpretation, to the deposit of results, publication of findings, and assessment of impact (**Larsen & Riis, 2012**;**MacColl & Jubb, 2011**; **Vaughan et al., 2013**). Libraries are accordingly moving into areas such as funding opportunities and grant writing, ethics review, data curation and repository management, poster design and conference hosting (physical and virtual), journal and monograph publishing, bibliometric evaluation and impact assessment (**Adema & Schmidt, 2010**; **Carroll, 2011a**;**Crow et al., 2012**; **Delserone, Kelly, & Kempf, 2010**; **Drummond & Wartho, 2009**;**Furlough, 2010**; **Healy, 2010**; **Herther, 2009**; **Lyon, 2012**; **McBain, Culshaw, & Walkley Hall, 2013**; **Vaughan et al., 2013**), alongside their established roles in collection stewardship, resource procurement and information literacy. The development of support for the management of research data in line with national policy is a particular concern (**Carlson & Kneale, 2011**; **Gold, 2007**;**Lynch, 2008**; **Lyon, 2012**).

Commentators increasingly assert that libraries need to reposition from support service to professional/scholarly partner and to transform their liaison librarians into embedded experts recognized as academic associates (**Carlson & Kneale, 2011**; **Duranceau, 2008**; **Fonseca &

Viator, 2009; **Gold, 2007**; **Vandegrift & Varner, 2013**), so that they can add value through "higher end research support" (**Corrall et al., 2013, p. 638**). **Vaughan et al. (2013, p. 313)** argue that service presentation is an important factor here: "By basing the service model on the research lifecycle, the library is then poised to be a partner through the entire process, not just at the bookends of research". The proposed shift from the "service-and-support" approach to a partnership model is significant. **Posner (2013, p. 45)** argues that contemporary digital humanities projects "do not need*supporters* – they need *collaborators*", explaining that libraries need to provide both infrastructure (tools, servers, etc.) and "intellectual labor" (knowledgeable librarians). **Vandegrift and Varner (2013, pp. 69, 76)** similarly explain how partnering scholars in digital humanities should go beyond collection building to content creation, by "making 'stuff'" (e.g., web sites, digitized collections, new tools), describing the model as "deep collaboration".

Others have questioned whether libraries have sufficient expertise for such roles, particularly the technical understanding and domain knowledge needed for data management and curation in different disciplines (**Gold, 2007**; **Lynch, 2008**). However, feedback from biomedical and science librarians collaborating as partners in a large-scale cross-disciplinary multi- institutional research project reveals that personal attitudes and interpersonal skills are the key factors for successful participation in team science, with the willingness to learn and adapt more important than a deep understanding of the technology (**García-Milian et al., 2013**). Sector organizations including Research Libraries UK, the Joint Information Systems Committee, and the Association of Research Libraries (ARL), have sponsored projects examining the skillsets needed for new or extended subject liaison and specialist roles (**Auckland, 2012**; **Cox, Verbaan, & Sen, 2012**; **Jaguszewski & Williams, 2013**). Library and information schools are also responding with modules, courses, specializations, internships, and institutes to prepare new professionals and update the existing workforce (**Creamer, Morales, Kafel, Crespo, & Martin, 2012**; **Harris-**

Pierce & Liu, 2012;**Lee & Tibbo, 2011**; **Stanton et al., 2011**; **Yakel, Conway, Hedstrom, & Wallace, 2011**).

The model of the librarian as a "scholar-practitioner" and active researcher is being promoted in several institutions (**Giesecke, 2010**; **McBain et al., 2013**;**Schrader, Shiri, & Williamson, 2012**), and being used to bring people with PhDs but no library degrees into research support roles in "alt/ac" (alternative to the academy) positions, particularly in relation to digital humanities developments (**Brunner, 2010**; **Giesecke, 2010**; **Nowviskie, 2013**; **Shore, 2012**). Even so, **MacColl and Jubb (2011, p. 10)** point out that despite recognizing their need for support, "researchers have little interest in the support services libraries have built for them". **Jahnke and Asher (2012, p. 16)** similarly report that "Few researchers see the library as a partner". Two large-scale surveys of academic faculty conducted in the US and UK confirm this message, with only half of the US respondents and one-third of the UK sample rating the library role in supporting their research productivity as "very important", placing this role as the least important of the six service areas specified (**Housewright et al., 2013a**; **Housewright, Schonfeld & Wulfson, 2013b**).

Some research libraries are responding to the challenge by moving beyond alignment with academic departments or schools to focus provision of in-depth information advice at the more specialized and dynamic level of research groups and projects, particularly in the biomedical and health sciences area (**Carlson & Kneale, 2011**; **Federer, 2013**; **Greyson & Surette, 2013**). Another significant trend is the creation of specialist positions dedicated to research support, either exclusively or in addition to a subject librarian role (**Bradbury & Weightman, 2010**; **Young & Lund, 2008**), including positions coordinating support for graduate students (**Covert-Vail & Collard, 2012**). The 2013 Ithaka survey of US library directors indicates that almost two-thirds of respondents in doctoral institutions expect to increase staffing for specialized faculty research support (**Long & Schonfeld, 2014**). Others have decided that traditional structures, such as the subject-specialist information consultant model, are no longer viable, and have reorganized their staffing to align with the broad institutional missions of research and teaching, or adopted business enterprise structures, with teams dedicated to service development and innovation (**Andrade & Zaghloul, 2010**;**Cotta-Schønberg, 2007**; **Covert-Vail & Collard, 2012**; **Drummond & Wartho, 2009**; **Franklin, 2009**; **Manchester University Library, 2012**; **Nutefall & Chadwell, 2012**).

1.2 Related work

Organization design has not received due attention in the debate on reshaping the library as research collaborator. Current literature on academic library structures consists largely of institutional case studies (**Andrade & Zaghloul, 2010**; **Chan & Soong, 2011**; **Franklin, 2009**; **Larsen & Riis, 2012**; **Nesdill, Love & Hunt, 2010**; **Quinlan & McHarg, 2012**; **Renfro & Neal, 2012**; **Shore, 2012**; **Warren, 2011**), with few comprehensive surveys. In North America, ARL last surveyed library organization in 1996, although the organization and staffing of specific elements of research support have been examined more recently, including digital humanities, research data management, and scholarly communication (**Bryson, Posner, St. Pierre, & Varner, 2011**; **Eustis & Kenney, 1996**; **Fearon, Gunia, Pralle, Lake & Sallans, 2013**; **Radom, Feltner-Reichert, & Stringer-Stanback, 2012**). In the UK, the convergence and merging of university libraries with information technology and other academic support services has dominated the discussion of library structures over the past two decades (**Bulpitt, 2012**; **Field, 2001**; **Hanson, 2005**; **Pugh, 1997a**).

Few writers on library organization relate their discussion to classic organizational theory, notable exceptions being **Pugh's (1997b, 2005, 2007**) efforts to establish a theoretical basis for the management of technology-driven change in libraries and **McGuigan's (2012)** essay identifying theories relevant to managing change in academic libraries.

Although library case studies rarely use terminology from organization design literature, it is evident that the dominant library model of recent decades is basically the mixed structure (**Child, 1988**), which organizes technical/back-office processes (e.g., acquisitions, cataloging) and generic front-of-house services (e.g., circulation) along functional lines and specialized client services in a divisional or market-based arrangement (i.e., individuals or teams devoted to supporting particular subjects, disciplines, departments, or schools). Many libraries have overlaid an element of matrix structure, by introducing cross-cutting roles to coordinate service provision across different groups (e.g., designating senior liaison librarians as lead specialist for key areas, such as scholarly communications); an emerging variant of this model uses functional specialists who do not have liaison assignments to supplement and support their liaisons (**Jaguszewski & Williams, 2013**).

1.3 Definition of key terms

Organization design has been defined as "the allocation of resources and people to a specified mission or purpose and the structuring of these resources to achieve the mission" (**Duncan, 1979**, p. 60). **Duncan (1979, p. 60)** also asserts that "Ideally, the organization is designed to fit its *environment* and to provide the information and *coordination* needed" (emphasis added). **Stanford (2013, p. 19)** offers an alternative definition of the concept emphasizing other dimensions:

"Arranging how to do the work necessary to effectively and efficiently achieve a business purpose and strategy whilst delivering high-quality customer and employee experience. Arranging involves aligning the organization with the strategy, creating coherent designs, while building trust among key stakeholders."

Design is thus conceived as the broader higher-level concept, with *structure*forming an important part of the design. However, the terms are often used synonymously, within both management and library literature. Another important point to note here is that "resources" in this context can include "tangible resources such as money and equipment and intangible resources such as knowledge" (**Miles, Snow, Fjeldstad, Miles, & Lettl, 2010, p. 93**).

1.4 Research question, aims and objectives

The central research question for the study is: *How are academic research libraries and their institutions organizing resources and services to support the research enterprise?*

The overall aims of the research are:

- To review organizational designs of research libraries and evaluate their fitness for purpose in the context of researcher support needs.
- The specific objectives of the project are:
- To review trends and developments in library support and services for research and researchers
- To examine design dimensions and structural features of a sample of research library organizations
- To identify specific elements of library organization of particular relevance in supporting research
- To analyze strengths and weaknesses of potential organizational models.

2. METHODOLOGY

This research uses an exploratory sequential mixed methods design (**Creswell & Plano Clark, 2011**) to survey organization structures of research libraries in the specific context of library service responses to challenges presented by the contemporary e-research environment. The purpose was first to qualitatively explore dimensions of library structural designs with a small single-country sample and then to quantitatively determine the extent to which the qualitative findings generalized to a large international sample. The present paper reports the results from the first phase only, but describes the complete design to explain the rationale for the initial research.

In the first (instrument-development) phase, data were collected via institutional websites, organization charts, strategy documents, job descriptions and other public-domain sources from a purposive sample of 24 universities in the UK. The Russell Group of leading UK research universities was used as the sampling frame.1 The use of websites and web-based documents as sources of data and objects of analysis has been recognized as an appropriate method of investigation for contemporary researchers, particularly in relation to organizational research (**Bryman, 2004**). Previous studies have used this type of Internet-based research to investigate library strategy documents, either as the sole method of data collection (**Corrall, 2007**; **Pacios, 2004**), or as the first phase of a multi-method study (**Aldrich, 2007**; **Corrall, 2008**; **McNicol, 2005**).

Both deductive and inductive strategies were used in the initial phase. A review of related literature informed the systematic browsing and keyword searching of websites, but the key outputs of the research (such as the categorizations of structural arrangements) were generated from the data analysis. Evidence was captured and recorded electronically as raw data (document extracts or whole documents), notes and initial codes. The data were analyzed using matrix analysis techniques, using color coding to highlight examples with striking features, which were then re-examined to select cases illustrating sharp contrasts (**Corrall, 2007**). Data matrices are particularly suited to making large amounts of qualitative data accessible and meaningful, enabling insights to emerge, and supporting both cross-site and within-site comparisons (**Miles & Huberman, 1994**; **Nadin & Cassell, 2004**).

The findings from this initial exploration will be used to develop assessment measures that can be administered to a large sample. In the planned second phase, data on service configurations, reporting relationships, divisions and specializations, coordination and integration will be collected from research library directors in Australia, New Zealand, North America, and selected European countries, using an online questionnaire survey. The reason for collecting qualitative data initially was that there is little guiding theory on the subject to inform the development of a survey instrument (**Creswell & Plano Clark, 2011, p. 86**).

3. RESULTS

Formal organization structure charts were found via web searches in eight cases, i.e., for one-third of the sample investigated (*n*=24). However, relevant dimensions and salient features of the structures in use at other sites were identified by browsing their web pages and consulting staff directories, service descriptions and other information available (including structure charts for the institution in some cases, and further particulars for job announcements in a few instances). The main elements examined were service configurations, reporting lines, internal organization, specialist positions, and coordination mechanisms. Throughout the analysis, the focus was specifically on elements of library organization of particular relevance in supporting research.

The headline findings from the data are as follows:

- Integrated library and computing services have fallen from favour

- Reporting lines for library and information services have shifted
- Library structures continue to mix function and market/subject elements
- New senior positions and titles are signalling strategic priorities
- Subject liaison arrangements are being enhanced and remodelled
- Specialist research positions are emerging as a significant trend
- Dedicated spaces for researchers are improving service visibility
- Groups and websites are promoting boundary-spanning services.

3.1 Service configurations

One-third of the services featured some form of convergence between the library and one or more cognate professional services, most commonly the computing/information technology (IT) service, which often also includes learning/education technology. This finding contrasts with the position a decade ago when more than half of the 20 institutions then in the Russell Group were converged and one-quarter had merged structures with fully integrated library and IT provision for several core functions (e.g., help desk, training). Among the currently converged organizations, four had retained the same basic structure for a decade or more, of which one was still a merged service, two were converged, and one was federated (separate organizations within an information services division of professional services).

Two organizations had deconverged their library and IT services and combined the library with other cognate services to form an academic services directorate and a student services directorate respectively. Two other institutions had formed newly converged services in the last five years, in both cases bringing the library and IT together, into an information directorate in one case, and academic services in the other (which also included student services and quality enhancement). However, the key finding here is that most members of the group now have separate library organizations, which includes four high-profile services that have deconverged their merged or converged service organization within the last three years (King's College London, the London School of Economics (LSE), and the Universities of Bristol and Nottingham).

The variety of service configurations was a feature of UK converged information service organizations in the 1990s and 2000s, and this pattern has continued as services have de-converged and re-converged more recently. For example, at both the University of Birmingham and the University of Exeter, Library services is one division within an Academic Services directorate of seven divisions, but the composition of each directorate and structure of the library division varies. Table 1 shows the breakdown (based on data collected in summer 2013).

Table 1. Sample composition of academic services directorates

University of Birmingham		University of Exeter	
Academic Services Divisions	*Library Services Departments*	*Academic Services Divisions*	*Library & Culture Services Groups*
• Birmingham Foundation Academy	• Academic Liaison, Learning, Research Support & Academic Skills	• Academic Support Unit (Cornwall)	*Library & Research Support Teams*
• Centre for Learning & Academic Development and Learning Spaces	• Collection Management & Development	• Education Quality & Enhancement	• Academic Engagement
• Student Employability	• Library Customer Support	• Employability & Graduate Development	• E-Strategy and Resource Development
• Library Services		• Exeter IT	• Heritage Collections
• Special Collections		• Library & Culture Services	• Library Customer Services
• Registry		• Planning & Resources	*Arts & Culture Team*
• Student Services		• Student Services	

At divisional level, in addition to library services, common functions here include academic development/educational enhancement, student employability (careers services), and student services/registry functions (combined at Exeter). However, technology services (including administrative systems, academic computing and student IT services) are part of the converged service at Exeter, but at Birmingham, in the re-converged service, IT Services is a separate unit within Professional Services reporting to the Registrar & Secretary – unlike the former converged information service organization at Birmingham, where IT was a central part of the original unified structure. At the library divisional level, Table 1 shows that the primary split into four functional sections is similar in both institutions, namely: academic liaison/engagement; customer services/support; collection/resource development; and special/heritage collections. However, Birmingham has departed from established practice by making Special Collections a separate division (not reporting to the Director of Library Services), while Exeter has brought in Arts and Culture as an additional service group reporting to the Assistant Director of Academic Services/Head of Library and Culture Services.

The make-up of the academic support department/team within each library also differs slightly. The Birmingham department includes responsibility for the University's Academic Skills Centre (which provides support in areas such as academic writing, maths and stats, and study skills), in addition to academic liaison, and research support (including supporting the REF and research data management); whereas Academic Skills Development at Exeter is located within the Education Enhancement division, and the Library's Academic Engagement group includes separate teams specializing in digital content and library marketing, and open access and data curation, in addition to liaison support for teaching and research. (At Birmingham, responsibility for managing and supporting the institutional repository and open access is part of the Collection Development and Management department.) Both institutions use the traditional subject librarian model to carry out liaison activities: at the time of data collection, Birmingham had 15

Subject Advisors supporting 5 Colleges, while Exeter had 10 Subject/Liaison Librarians supporting 4 Colleges and separate Business and Medical Schools (part of the College structure at Birmingham). The numbers of liaisons are commensurate with the respective sizes of the populations served, Birmingham being approximately 50 percent larger than Exeter.

The Information Services directorates that have continued as merged, converged or federated services are now more uniform in their service portfolios. All include libraries (and archives), IT (academic computing and corporate systems), and learning technologies, usually including responsibility for both equipping and managing teaching spaces, but not covering academic skills (other than information literacy), which are typically provided by academic support/learning development units, nor other student services functions. Within the sample, although some instances of shared support functions were found (e.g., Operations and Central Projects at Birmingham; Strategic Support, Finance, HR, etc. at Cardiff; Planning and Resources at Exeter), only Edinburgh currently displays elements of the more radical merged service structures that were adopted to varying extents by Birmingham and other leading research universities from the mid 1990s.

The key features of the Edinburgh structure are the combined library, e-learning and IT first-line help/support service, and the organization of liaison librarians and IT professionals into three College-facing consultancy teams within the User Services Division. The consultancy teams are in a different part of the organization from other library staff, who are mainly located in the Library and Collections Division. This means they are separated from staff working in related research and learning support services, including teams and sections dealing with scholarly communication, and collection development and management, and also do not report to the *de facto* library service head, the Director of Library & University Collections.

3.2 Reporting relationships

In UK universities, the Librarian traditionally reported to the Vice-Chancellor or Principal (as the chief executive officer of the institution), which often differentiated the library from the IT/computing service; but the adoption of more formal management arrangements in higher education institutions over the past two decades and the reconfiguration of administrative support functions into groups of professional services has resulted in different structures. Nearly three-quarters of the library directors in the sample report directly (or indirectly, via another senior officer) to the University's Registrar/Secretary/Chief Operating Officer. Chief Operating Officer (COO) is becoming the dominant title in use, found in half of the institutions.

Only Oxford and Cambridge seem to have retained their traditional reporting line in its original form, although the head of the converged service at Edinburgh (who holds the formal title of Vice-Principal for Knowledge Management, CIO and University Librarian) reports directly to the Principal. At the time of data collection, the university librarian/library director at Manchester and Imperial reported to the equivalent of the Deputy Vice-Chancellor (Deputy Rector at Imperial); at Leeds and Nottingham, they reported to a Pro-Vice Chancellor (PVC), in both cases the PVC with responsibility for education/teaching and learning. The grouping of the library with education/student services is another notable trend in structural arrangements, reflected in the recently converged academic service model adopted at Exeter, the reconverged academic services directorate at Birmingham, and reconverged student services at Queen Mary (University of London). In addition, the deconverged library service at King's now reports to the College Secretary via the Director of Students & Education; similarly at Bristol, the reporting line is via the Deputy Registrar, responsible for "the student-facing and education-related services".

It is debatable whether a PVC reporting line is preferable to the COO line, particularly where the PVC portfolio has a focus on teaching/students. Reporting directly to the COO role can position the library director alongside other professional service heads, including both student services and research services, which is particularly valuable in the context of collaborative service development with the research office or equivalent in areas such as managing research data and measuring research impact.

3.3 Internal organization

The size and shape of the library superstructures vary significantly, with leadership teams ranging from three to eight members. The titles of senior management positions and labels used for the main divisions/departments/operational units also vary, and some libraries combine one or more functions together, but overall there was a striking consistency in the way activities were grouped into the following five areas:

- Academic services/research and learning support (generally subdivided by subject field)
- Collection development/information resource management
- Special collections and archives/heritage collections
- Customer services/reader services/user support
- E-strategy/digital services/information technologies.

The academic (learning and research) liaison/support function was explicitly represented in the senior team in three-quarters of the sample. In one-third of these (six cases), the relevant manager had other responsibilities (e.g., Planning & Resources, Collection Development/Management, Special Collections, User Services, Service Development). In one novel example, at the University of Manchester, research services, teaching and learning support, and academic engagement were distributed among three different members of the leadership team, and combined with other related responsibilities. At Edinburgh, the merged Information Services organization separates consultancy including library liaison services (in the User Services Division) from other library-based research and learning services (in the Library & Collections Division).

Academic Services was the descriptor used most consistently, found in the job titles of seven senior managers. However Research, or Research & Learning (or Learning & Research) featured in seven titles of associate/assistant directors/heads of division or equivalent. The word Liaison was included in four titles, and the term Academic Engagement in two jobs. Other variants included the position of Head of Relationship Management at York, and the five Heads of Faculty Services in the senior team at Southampton (where academic liaison is directed by the Deputy Librarian). The most complex title identified was Assistant Director – Academic Liaison, Learning & Research Support, and Academic Skills at the University of Birmingham (a position that has subsequently been changed to Academic Engagement).

The collection development/management function was equally prominent, also being represented in the senior teams of three-quarters of the sample. However, the term "collection" was used explicitly in slightly less than half the sample, and the titles of the relevant roles, divisions, or sections also varied within these cases, e.g., Collection & Access Management (Nottingham), Collection Development & Description (Cambridge), Collection Services (Durham, Leeds). The term "resources" seems to be gaining currency as an alternative to "collection", suggesting the concept of the library collection may become less prominent as migration from purchased printed materials to licensed e-journals and e-books continues. At York, the relevant senior position is Head of Content.

In approximately half of the cases featuring a collection development or management role in the top team, the collections function was combined with the management or strategic development of library systems (in two cases along with facilities), represented by roles such as Head of Collections and eStrategy (Sheffield – a position that has now evolved into Assistant Director for Academic Services by adding responsibility for the Learning & Research Services team to the Procurement & Content and Digital Services & Systems teams), Head of eStrategy & Resource Development (Exeter – a position that is now combined with academic liaison responsibilities in an enlarged Academic Support portfolio), and Head of Information Management & Library Development (Bristol). A slightly less prevalent, but equally significant, trend (found in around one-quarter of the total sample), was the creation of senior roles solely concerned with the library's information systems/information technologies portfolio, notable examples here including: Associate Director, Digital Library Programmes & Information Technologies (Oxford), Head of Digital Library Services (Leeds), and Head of Digital Technologies & Services (Manchester).

Around half of the institutions had special collections and/or archives represented as a distinct role in the library's senior team at the level of divisional/departmental head, including two who used the alternative title of Head of Heritage Collections (Durham and Exeter). Also in around half of the sample, delivery of frontline services to users (such as lending/circulation and enquiry/help services) was represented explicitly in the senior group, e.g., Assistant Director – Customer Services (Kings), Head of Library Customer Services (Exeter). In several other cases, customer/user services was part of a larger portfolio for a member of the executive. Customer Services (or Customer Support) was by far the most common label used for this function, followed by four instances of User Services, two of Reader Services, and only one where the term Public Services was used.

3.4 Subject liaison

The importance of subject-based academic support was evident in the structures of all 24 libraries in the sample. With the exception of the

Universities of Cambridge (where the library's strategic objectives include development of "integrated, subject-based provision for teaching and learning as well as research") and Manchester (where academic liaison through teams of faculty librarians was recently discontinued), every institution currently uses some form of subject librarian arrangement to provide specialized support to students and academic/research staff. Seven use the title *Subject* Librarian (or Subject Advisor or Subject Consultant) for this role, while 11 use *Liaison* Librarian (or Academic Liaison Librarian, or Faculty Liaison Librarian). (Other titles found include Academic Support Librarian, College Team/Support Librarian, Faculty Team Librarian, and Information Specialist.) In ten cases, the library had a team structure mirroring the institutional structure of Colleges or Faculties, with subject/liaison librarians led by Heads of Faculty Services, Faculty Team Leaders, Faculty Librarians, or similar. Some libraries had additional structural differentiation by including *Assistant* Liaison Librarians, or Liaison *Assistants* in their teams.

The strength of liaison teams/subject support varied across the sample from six liaison librarians at Queen Mary (University of London) and eight at Durham, to 32 subject librarians at Cardiff and 52 at Oxford (at the time of data collection). Although generally the liaison librarian title was associated with smaller-scale provision, this was not universally the case, as the Queen's (Belfast) structure included only eight Subject Librarians (led by three Faculty Librarians), whereas Southampton had 24 Liaison Librarians, led by five Heads of Faculty Services (who also fulfilled liaison roles). The mean number across the sample was 15.9 and the median was 13.5. However, these figures should be seen as approximations only. They do not always represent the total number of librarians involved in subject-based learning support and research services, as many libraries have recently established specialist positions that focus solely on learning or research support, or a particular aspect of academic support. In addition, the method used to collect data on subject liaisons did not differentiate between full-time and part-time liaison staff.

3.5 Specialist positions

As reported in the Introduction, support for research has traditionally been a formal part of the subject/liaison/faculty librarian role, alongside support for teaching and learning, but published literature indicates a trend towards libraries putting more effort into the delivery and promotion of support for researchers, having identified opportunities to provide more specialized "higher-end" services, and also become aware that researchers often have a limited view of what librarians provide (tending to focus on resource procurement and collection management). A significant development in the past decade has thus been the creation of new roles within the academic services/support area designed to provide specialist support for researchers in areas such as literature searching, systematic reviews, current awareness, reference management, citation analysis, publishing advice, open access and data curation.

One-third of the sample libraries had established a specialist position with a research support focus, mostly within the past five years. In some examples, the role is similar to and positioned alongside existing subject/liaison librarians, with titles such as Academic Liaison Librarian (Researcher Support) at Durham, or Research Support Librarian at Queen Mary. In most cases, the post-holder is also expected to fulfill a management coordination or strategic leadership role, both carrying out specialist work and providing direction for research-related service developments across library teams; examples of relevant job titles here include Research Support Leader at Liverpool, Library Liaison Manager (Research) at King's, Head of Scholarly Communications at Southampton, and Academic Support Manager (Research) at Warwick – where the post-holder is supported by a dedicated Academic Support Librarian. In a few instances, the post-holders also act as subject librarians/academic liaison contacts (e.g., Senior Consultant, Research Support at Cardiff, and Research Support Services Manager at LSE).

In addition to this type of cross-cutting coordination and development role, several libraries have established more specific research support positions, whose remits have evolved in line with additional needs identified. For example, the Research Skills & Development Librarian at Cambridge directs the University Library's Research Skills Programme, but is also a member of the University's Open Access Project team; at Imperial, the intended focus of the new position of Research Support Librarian (Medicine) was mediated literature searching for systematic reviews, but the role developed to provide support in other areas, such as data management. Other longer established examples of specialist research support provision in libraries include two largely grant-funded units, the Support Unit for Research Evidence (SURE, established in 2000) at Cardiff, which specializes in systematic reviews of the literature; and the Centre for Research Communications at Nottingham (established formally in 2009, but with a history going back to the SHERPA project of 2002–2006), which carries out research and development work in scholarly communication, open access and related areas.

Management of the institutional repository/research archive is now an established library function, but its location within the organization structure varies. In two instances within the sample, the repository function had been subsumed within the institutional research information system managed by the research office; in another two examples, the repository service was a joint venture with the university computing service, but located in the library. However, repository management was mostly a library responsibility, generally located either in the library's systems/IT/digital services division (10 examples), or with academic/research and learning services (9 examples). In two cases the research support librarian or equivalent was formally designated as the institutional repository manager, and in another two cases, the research support leader had formerly filled the role of repository manager. At Birmingham, the repository is the responsibility of the Collection Development and Management department, where the Digital Assets team is located (in the absence of a library IT department).

Policy developments and funding mandates related to research assessment, open access and data management have resulted in many new and redesigned specialist library positions and teams to manage and coordinate institutional support in these areas. For example, Edinburgh has a Scholarly Communications Team of four people providing support for policy development and implementation, open access advice and advocacy, intellectual property rights, bibliometrics, repository administration, publications submission, and metadata quality. Exeter has an Open Access and Data Curation Team, comprising a Manager, a Data Curation Officer, and an Open Access Administrator. At Bristol, an institutional Research Data Service was launched in summer 2013, as a Library-led collaboration with IT Services and Research & Enterprise Development; it is staffed by a Manager, two Senior Research Data Librarians and an Assistant Research Data Librarian. Sheffield has formed a new Research Support Team within its Learning & Research Services group, which includes two Coordinators for Open Access and Research Data Management, two Open Access Assistants and a Research Data Management Liaison Assistant.

3.6 Dedicated spaces

The University of Warwick Library launched its Wolfson Research Exchange in 2008 as an innovative suite of specialist services and dedicated technology-rich spaces for researchers, with provision for quiet study, interdisciplinary networking, academic meetings, and social activities, and a programme of events on topics such as funding opportunities, grant applications, data management, journal impact, and spinout companies, "offered not only by the library but by other university departments supporting research too" (**Carroll, 2011b, p. 55**). Novel services include Research Match ("a unique service which profiles your research and makes it visible") and Collaborative Sandpits (interactive multi-day workshops to create project proposals for funding).

With separate programming for graduate research students, early career researchers, and academics/research staff, the facility has its own dedicated staffing, with a Research Exchange and PG Hub Manager, a Coordinator, Support Officers, and Assistants, who work alongside other managers and academic support specialists in the Academic Services division of the Library. *In situ*support is provided by a team of part-time Advisers, who are all current or recent PhD students (**Carroll, 2011a**). The Library website presents the facility as "A space, a website, a community", reflecting its several dimensions, including its hosting of networking evenings, interdisciplinary special interest groups, online peer to peer discussions, and blogs.

The development of dedicated purpose-designed space for researchers emerges as a significant trend creating physical presence for renewed efforts by libraries to position themselves as research partners and collaborators. The new research spaces typically have card-controlled entry to restrict access and demonstrate that libraries understand that researchers want and need physical facilities that are separate from undergraduates. Queen Mary opened a new 58-seater Research Reading Room for graduate researchers and academic staff in 2011, after a consultation identified the need for a quiet, comfortable working environment away from busier noisier parts of the Library. The facility was "designed along similar lines to The British Library Reading Rooms, with high specification study desks finished in oak and leather, providing larger than standard worktop space, and LED task lighting at each desk". Birmingham, similarly, has a Researchers' Room "reserved for use by postgraduate research students and academics only".

Social space and technology provision generally differentiate contemporary researcher facilities from traditional reading rooms. For example, Durham is developing a researchers' study area for the exclusive use of postgraduate research students, with individual, collaborative and social areas, which is also being used to promote information about library support for research. At Exeter, the Old Library housing the University's special collections has been refurbished and renamed the Research Commons, with a new large reading room, enhanced IT, multi-media seminar room, and café-style break out space and terrace area. At York, where the Library was able to extend into an adjoining building, renovation has created new improved research space, including a study area, specifically for researchers; an informal research lounge area; and – the most unusual marketing feature – a Research Hotel, a suite of bookable rooms that visiting academic staff or researchers can reserve for their exclusive use as temporary work areas. The website explains that "The hotel concept does not extend to the provision of sleeping arrangements, but does offer use of the research lounge".

3.7 Boundary-Spanning Activities

Policy developments in the UK, notably the Engineering and Physical Sciences Research Council Policy Framework on Research Data (**EPSRC, 2011**) and the Research Councils UK (**RCUK, 2013**) Policy on Open Access have given rise to other structural arrangements, such as steering/working groups bringing institutional stakeholders together to facilitate compliance with funder requirements. Such groups are typically led or chaired by a Pro-Vice Chancellor for Research or senior research administrator and usually include representatives of academic schools/faculties/colleges in addition to key staff from the library, IT service and research office or equivalent. For example, at Cambridge, an Open Access Working Group was set up in 2012 to draft an institutional policy and make recommendations on implementation, followed by an Open Access Project Board in 2013 "to manage the transition to Open Access". At Exeter, "an Open Access and Research Data Management Policy Task and Finish Group was set up in March 2012 to develop an institutional policy on research data and Open Access issues". Leeds, University College London (UCL) and Warwick all have a two-tier structure for managing research data, with a high-level steering/oversight group or project board complemented by a Research Data Working Group at Leeds, a Technical & Operations Group at UCL, and an Action/Implementation Group at Warwick; the UCL structure also includes a User Group, which meets less frequently.

Institutional websites offer another opportunity to coordinate, focus and promote services to support research and researchers. Three-quarters of the library and information services in the sample had a set of web pages bringing together information targeted at researchers. At one-third of the sites, there was a prominent link from the library/information services home page, labeled Research Support, Researchers, or simply Research; at another eight, users could reach the relevant page with two clicks. The web pages varied in the topics featured, but there was a clear trend towards selectivity and integration, bringing together resources judged most useful to researchers, offered both by the library and other institutional units (such as research computing/IT services and the research office), and also trying to present the information from a research user viewpoint, rather than a service provider perspective. For example, at King's College London, the Library Research Support home page has the sub-title "Support through the research lifecycle" and a choice of six categories to click: Ideas, Research funding, Managing information, Disseminating research, Evaluating research, and Preserving research; there is also a link to the Library Research Support site from the institutional Research Support website. The Researcher@Library website at Leeds "gives researchers at the University of Leeds a single point of access to information, support and resources across the University".

Research impact, open access, and data management feature strongly on these web pages, reflecting current preoccupations of the UK research community, and also demonstrating how libraries are taking advantage of recent policy developments to position themselves as central to the research process. In almost all institutions in the sample, the library has been assigned responsibility for managing the funds allocated for Article Processing Charges incurred in compliance with funder requirements for gold open access. Most libraries have developed new web pages to support this process, typically including tables and/or decision trees/flow charts adapted to show how the institution is interpreting and implementing the relevant policies. Web pages supporting research data management were found at two-thirds of the institutions, with half of the sites clearly branded as part of the library/information services offer. Five services (including two projects led by the library) were presented as university-level offerings, usually acknowledging contributions by different services, and two services were based in the IT service. The websites were at various stages of development, with two consisting solely of links to external sites.

4. DISCUSSION AND CONCLUSIONS

4.1 Convergence

The move away from the converged service model is one of the striking findings of the study. However, the distinction between "organizational" and "operational" convergence (**Field, 2001**, p. 268) is important here. *Organizational*convergence (i.e., formally combining two or more services to form one organization) is evidently in decline within the sample, especially structural designs merging libraries with computing/IT services. *Operational* convergence (i.e., separate services/departments collaborating to coordinate their activities to improve conference and effectiveness), in contrast, is arguably more prevalent than ever, with libraries extending and deepening their collaborations and partnerships beyond IT and educational development colleagues to other professional services, such as research offices. This is particularly evident in:

- services (e.g., institutional repositories, research data services) delivered as collaborative ventures with IT services and/or research offices

- relationships and structures (e.g., committees and groups) supporting boundary-spanning activities in areas such as data management and open access
- unified research support websites providing a single point of access to university-wide resources for researchers, which could be described as an example of a *virtually* converged information service.

4.2 Positioning

The location of the library/information service organization in the institutional organizational structure is rarely discussed in the literature, but is important because it has implications for the allocation of resources, the coordination of activities, and the perception of the service by stakeholders. The two significant trends here were first, the administrative positioning of the library within professional services reporting to the Chief Operating Officer or equivalent, which was identified as the dominant model; and secondly, an increasing tendency to group libraries with "student-facing and education-related services" through reporting lines and directorate structures. The latter is arguably not a positive move for libraries trying to reposition themselves as credible collaborators and partners in research activities.

4.3 Specialization

The survey showed that departmentalization within libraries has evolved beyond the traditional reader/technical services model to a focus on five strategic areas of activity: information resources, academic engagement, customer service, heritage collections, and digital technologies, although there was significant variation in the size of leadership teams. Use of the term "Research" in the job titles of around one-third of the senior management positions identified sends a signal to stakeholders about the priority attached to this area. The strong focus on information systems/technologies identified at the senior management level is not surprising, but it was interesting to find that in many cases the technology portfolio was combined with another area of responsibility, and most frequently with information resources/ collections, which in effect continued the tradition of locating systems librarians in a technical services division/department.

Specialization of library professional staff continues to focus on the subjects/academic disciplines taught and researched in the institution, but subject/liaison roles are increasingly being complemented and supplemented by coordinating roles (such as research support managers) and functional specialists for particular areas of activity (e.g., research data management, open access publishing). The hybrid approach to specialization found in UK research libraries echoes structural developments reported in North America (**Covert-Vail & Collard, 2012**; **Jaguszewski & Williams, 2013**).

An interesting dimension of specialization is the growing trend to extend the development of specialized research support into dedicated spaces for researchers. The space-as-service concept is gaining currency internationally as a visible contribution to research, particularly in support of early career researchers, and interdisciplinary collaboration, with the research commons beginning to emerge as a parallel development to the successful learning commons movement (**Corrall & Lester, 2013**; **Daniels, Darch, & de Jager, 2010**;**Ohio State University Libraries, 2012**). UK examples range from traditional reading rooms to the fully-serviced technology-rich space at Warwick, using early career researchers as peer support advisors, and offering novel services such as research profile matching and suites of bookable rooms for individuals and meetings (**Carroll, 2011a,b**).

4.4 Coordination

Coordination within libraries is being formalized through roles such as research support managers and leaders, and data management and open access coordinators (and also by internal committees and working groups). Coordination with other services and operations is being pursued through and facilitated by institution-wide groups at both strategic and operational level, typically established to deal with external mandates, but offering the potential of additional benefits for librarians in relationship building, professional credibility, and institutional positioning. As noted above, websites are an additional valuable way of coordinating research services to provide a single point of access for researchers and can be used not only to pull together library and other institutional resources and services, but again to position the library as a major player in the research arena. Research support websites complement the work of research support coordinators in libraries, but can also fulfill that role in libraries which do not have an individual tasked with that responsibility.

4.5 Outcomes

The purpose of the present study was to explore qualitatively how a sample of academic research libraries are organizing resources and services to support their research mission with the intention of identifying important dimensions of organizational design that could be used to develop an assessment instrument for a larger scale international survey. The study achieved its objectives and the results both confirm and extend the findings of related research investigating library plans, roles and skills for supporting and serving researchers. Although the sample was limited to 24 libraries in one country, the results identify structural variables related to library support for research that can be used as a provisional framework for libraries in the UK and other countries to examine and evaluate their own structures.

The present study was conceived as the first phase of a larger scale international survey of organizational design in research libraries and the findings are being used to develop a questionnaire to collect quantitative data on important structural and contextual variables identified from this exploratory study and ongoing review of related literature. The dimensions to be investigated include:

- positioning of libraries within their parent institutions
- scope of library/information service organizations
- size and focus of library leadership teams
- specialization of units, roles and space
- formal and informal convergence of activities
- mechanisms facilitating boundary-spanning activities.

ACKNOWLEDGMENTS

A summary of this study was presented at the 2013 LIBER Annual Conference in Munich, Germany. A synopsis of the key findings was also published in *Library Connect Newsletter, 11*(3), The Library and the Research Enterprise (pp. 3–4).

NOTES

[1]http://www.russellgroup.ac.uk.

REFERENCES

1. Adema, J., & Schmidt, B. (2010). From service providers to content producers: New opportunities for libraries in collaborative open access book publishing. New Review of Academic Librarianship, 16(sup. 1), 28–43.

2. Aldrich, A.W. (2007). Following the phosphorous trail of research library mission statements into present and future harbors. ACRL 13th National Conference, March 29–April 1, 2007, Baltimore, MD (pp. 304–316). Chicago, IL: Association of College & Research Libraries. Retrieved April 30, 2014, from http://www.ala. org/ acrl/sites/ala.org.acrl/files/content/conferences/confsandpreconfs/ national/baltimore/papers/304.pdf

3. Andrade, R., & Zaghloul, R. (2010). Restructuring liaison librarian teams at the University of Arizona Libraries, 2007–2009. New Library World, 111(7/8), 273–286.

4. Auckland, M. (2012). Re-skilling for research: An investigation into the role and skills of subject and liaison librarians required to effectively support the evolving information needs of researchers. London: RLUK Research Libraries UK. Retrieved April 30, 2014, from http://www.rluk.ac.uk/wp-content/uploads/2014/02/RLUK-Re-skilling.pdf

5. Bourg, C., Coleman, R., & Erway, R. (2009). Support for the research process: An academic library manifesto. Dublin, OH: OCLC Research. Retrieved April 30, 2014, from http://www.oclc.org/ content/ dam/research/publications/library/2009/2009-07.pdf

6. Bradbury, K., & Weightman, A. (2010). Research support at Cardiff University Library. SCONUL Focus, 50, 65–70. Retrieved April 30, 2014, from http://www.sconul.ac.uk/sites/default/ files/ documents/19_2.pdf

7. Brunner, M.L. (2010). PhD holders in the academic library: The CLIR Postdoctoral Fellowship Program. In K. Williams & S. Walter (Eds.), The expert library: Staffing, sustaining, and advancing the academic library in the 21st century (pp. 158–189). Chicago, IL: Association of

College & Research Libraries. Retrieved June 23, 2014, from http://escholarship.org/uc/item/05j228r4

8. Bryman, A. (2004). Social research methods. (2nd. ed.). Oxford: Oxford University Press

9. Bryson, T., Posner, M., St. Pierre, A., & Varner, S. (2011). Digital humanities: SPEC Kit 326. Washington, DC: Association of Research Libraries

10. Bulpitt, G. (Ed.) (2012). Leading the student experience: Super-convergence of organisation, structure and business processes. London: Leadership Foundation for Higher Education. Retrieved April 30, 2014, from http://www.lfhe.ac.uk/en/ components/ publication.cfm/S3%20-%2005

11. Carlson, J., & Kneale, R. (2011). Embedded librarianship in the research context: Navigating new waters. College & Research Libraries News, 72(3), 167–170. Retrieved April 30, 2014, from http://crln.acrl.org/content/72/3/167.full

12. Carroll, D.L. (2011a). Fostering a community of scholars at the University of Warwick: The Wolfson Research Exchange. New Review of Academic Librarianship, 17(1), 78–95. doi: 10.1080/ 13614533.2011.547820. Retrieved June 23, 2014, from http://wrap.warwick.ac.uk/4540

13. Carroll, D. (2011b). What researchers want: A personal case study. SCONUL Focus, 51, 52–56. Retrieved April 30, 2014, from http:// www.sconul.ac.uk/sites/default/files/documents/16_3.pdf

14. Chan, D.L.H., & Soong, S.C. (2011). Strategic repositioning in a dynamic environment. Library Management, 32(1/2), 22–36.

15. Child, J. (1988). Organization: A guide to problems and practice (2nd ed.). London: Paul Chapman

16. Corrall, S. (2007). Benchmarking strategic engagement with information literacy in higher education: Towards a working model. Information Research, 12(4), paper 328. Retrieved April 30, 2014, from http://www.informationr.net/ir/12-4/paper328.html

17. Corrall, S. (2008). Information literacy strategy development in higher education: An exploratory study. International Journal of Information Management, 28(1), 26–37.

18. Corrall, S., & Lester, R. (2013). The researcher's view: Context is critical. In L. Watson (Ed.), Better library and learning spaces: Projects, trends and ideas (pp. 183–192). London: Facet

19. Corrall, S., Kennan, M.A., & Afzal, W. (2013). Bibliometrics and research data management services: Emerging trends in library support for research. Library Trends, 61(3), 636–674. Retrieved April 30, 2014, from http://d-scholarship.pitt.edu/18948/.

20. Cotta-Schønberg, M. (2007). The changing role of the subject specialist. Liber Quarterly, 17(3/4). Retrieved April 30, 2014, from http://liber.library.uu.nl/index.php/lq/article/view/7890/8111

21. Covert-Vail, L., & Collard, S. (2012). New roles for new times: Research library services for graduate students. Washington, DC: Association of Research Libraries. Retrieved April 30, 2014, from http://www.arl.org/storage/ documents/ publications/nrnt-grad-roles-20dec12.pdf

22. Cox, A., Verbaan, E., & Sen, B. (2012). Upskilling liaison librarians for research data management. Ariadne, 70. Retrieved April 30, 2014, from http://www.ariadne.ac.uk/issue70/cox-et-al

23. Creamer, A.T., Morales, M.E., Kafel, D., Crespo, J., & Martin, E.R. (2012). A sample of research data curation and management courses. Journal of eScience Librarianship, 1(2), 88–96. doi: 10.7191/jeslib.2012.1016. Retrieved April 30, 2014, from http://escholarship.umassmed.edu/jeslib/vol1/iss2/4/

24. Creswell, J.W., & Plano Clark, V.L. (2011). Designing and conducting mixed methods research (2nd ed.). Thousand Oaks, CA: SAGE

25. Crow, R., Ivins, O., Mower, A., Nesdill, D, Newton, M., Speer, J., & Watkinson, C. (2012). Library publishing services: Strategies for success. Washington, DC: SPARC. Retrieved April 30, 2014, from http://wp.sparc.arl.org/lps/

26. Daniels, W., Darch, C., & de Jager, K. (2010). The research commons: a new creature in the library. Performance Measurement and Metrics, 11(2), 116–130.

27. Delserone, L.M., Kelly, J.A., & Kempf, J.L. (2010). Connecting researchers with funding opportunities: A joint effort of the Libraries and the University Research Office. Collaborative Librarianship, 2(1), 33–37. Retrieved April 30, 2014, from http://collaborativelibrarianship.org/index.php/jocl/article/view/59/39

28. Drummond, R., & Wartho, R. (2009). RIMS: The Research Impact Measurement Service at the University of New South Wales. Australian Academic & Research Libraries, 40(2), 76–87.

29. Duncan, R. (1979). What is the right organization structure? Decision tree analysis provides the answer. Organizational Dynamics, 7(3), 59–80

30. Duranceau, E.F. (2008). The "Wealth of Networks" and institutional repositories: MIT, DSpace, and the future of the scholarly commons. Library Trends, 57(2), 244–261.

31. EPSRC (2011). EPSRC policy framework on research data. Swindon: Engineering and Physical Sciences Research Council. Retrieved April 30, 2014, from http://www.epsrc.ac.uk/about/standards/researchdata/Pages/policyframework.aspx

32. Eustis, J.D., & Kenney, D.J. (1996). Library reorganization & restructuring: A SPEC kit. Washington, DC: Association of Research Libraries

33. Fearon, D., Gunia, B., Pralle, B.E., Lake, S., & Sallans, A.L. (2013). Research data management services: SPEC Kit 334. Washington, DC: Association of Research Libraries

34. Federer, L. (2013). The librarian as research informationist: A case study. Journal of the Medical Library Association, 101(4), 298–302.

35. Field, C.D. (2001). Theory and practice: Reflections on convergence in United Kingdom universities. LIBER Quarterly, 11(3), 267–289.

Retrieved April 30, 2014, from http://liber.library.uu.nl/index.php/lq/article/view/7646/7682

36. Franklin, B. (2009). Aligning library strategy and structure with the campus academic plan: A case study. Journal of Library Administration, 49(5), 495–505.

37. Fonseca, A.J., & Viator, V.P. (2009). Escaping the island of lost faculty: Collaboration as a means of visibility. Collaborative Librarianship, 1(3), 81–90.

38. Furlough, M.J. (2010). The publisher in the library. In S. Walter & K. Williams (Eds.). The expert library: Staffing, sustaining, and advancing the academic library in the 21st century (pp. 190–233). Chicago, IL: Association of College and Research Libraries. Retrieved June 23, 2014

39. García-Milian, R., Norton, H.F., Auten, B., Davis, V.I., Holmes, K.L., Johnson, M., & Tennant, M.R. (2013). Librarians as part of cross-disciplinary, multi-institutional team projects: Experiences from the VIVO collaboration. Science & Technology Libraries, 32(2), 160–175. doi: 10.1080/0194262X.2013.791183. Retrieved June 23, 2014, from http://www.ncbi.nlm.nih.gov/pmc/articles/PMC3700548/

40. Giesecke, J. (2010). Finding the right metaphor: Restructuring, realigning, and repackaging today's research libraries. Journal of Library Administration, 51(1), 54–65.

41. Gold, A. (2007). Cyberinfrastructure, data, and libraries, Part 2. Libraries and the data challenge: Roles and actions for libraries. D-Lib Magazine, 13(9/10).

42. Greyson, D., & Surette, S. (2013). "You're just one of the group when you're embedded": report from a mixed-method investigation of the research-embedded health librarian experience. Journal of the Medical Library Association, 101(4), 287–297.

43. Hanson, T. (Ed.) (2005). Managing academic support services in university: The convergence experience. London: Facet

44. Harris-Pierce, R.L., & Liu, Y.Q. (2012). Is data curation education at library and information science schools in North America adequate? New Library World, 113(11/12), 598–613

45. Healy, A.M. (2010). Increasing the visibility of the library within the academic research enterprise. Issues in Science and Technology Librarianship, 63.

46. Herther, N.K. (2009). Research evaluation and citation analysis: key issues and implications. The Electronic Library, 27(3), 361–375.

47. Housewright, R., Schonfeld, R.C., & Wulfson, K. (2013a). Ithaka S+R US faculty survey 2012. New York: Ithaka S+R. Retrieved April 30, 2014, from http://www.sr.ithaka.org/research-publications/us-faculty-survey-2012

48. Housewright, R., Schonfeld, R.C., & Wulfson, K. (2013b). Ithaka S+R/JISC/RLUK US survey of academics 2012. New York: Ithaka S+R. Retrieved April 30, 2014, from http://sr.ithaka.org/research-publications/ ithaka-sr-jisc-rluk-uk-survey-academics-2012

49. Jahnke, L.M., & Asher, A. (2012). The problem of data: Data management and curation practices among university researchers. In L. Jahnke, A. Asher & S.D.C. Keralis, The problem of data (pp. 3–31). Washington, DC: Council on Library and Information Resources. Retrieved April 30, 2014, from http://www.clir.org/ pubs/ reports/ pub154

50. Jaguszewski, J.M., & Williams, K. (2013). New roles for new times: Transforming liaison roles in research libraries. Washington, DC: Association of Research Libraries. Retrieved April 30, 2014, from http://www.arl.org/storage/documents/publications/NRNT-Liaison-Roles-final.pdf

51. Larsen, S. & Riis, C. (2012). Aarhus University Library: Restructuring a university's library organisation. Bibliothek Forschung und Praxis, 36(2), 235–240.

52. Lee, C.A., & Tibbo, H. (2011). Where's the archivist in digital curation: Exploring the possibilities through a matrix of knowledge and skills. Archivaria, 72, 123–168.

53. Long, M.P., & Schonfeld, R.C. (2014). Ithaka S+R Library Survey 2013. New York: Ithaka.

54. Lynch, C. (2008). The institutional challenges of cyberinfrastructure and e-research. EDUCAUSE Review, 43(6), 74–88. Retrieved April 30, 2014

55. Lyon, L. (2012). The informatics transform: Re-engineering libraries for the data decade. International Journal of Digital Curation, 7(1), 126–138.

56. MacColl, J., & Jubb, M. (2011). Supporting research: Environments, administration and libraries. Dublin, OH: OCLC Research. Retrieved April 30, 2014, from http://www.oclc.org/resources/research/publications/library/2011/2011-10.pdf

57. Manchester University Library (2012). News from member libraries. SCONUL Focus,56, 61–62. Retrieved April 30, 2014, from http://www.sconul.ac.uk/sites/default/files/documents/20_6.pdf

58. McBain, I., Culshaw, H., & Walkley Hall, L. (2013). Establishing a culture of research practice in an academic library: an Australian case study. Library Management, 34(6/7), 448–461.

59. McGuigan, G.S. (2012). Addressing change in academic libraries: A review of classical organizational theory and implications for academic libraries. Library Philosophy and Practice, Paper 755. Retrieved April 30, 2014, from http://digitalcommons.unl.edu/libphilprac/755/

60. McNicol, S. (2005). The challenges of strategic planning in academic libraries. New Library World, 106(11/12), 496–509.

61. Miles, M.B. & Huberman, A.M. (1994). Qualitative data analysis: An expanded sourcebook. (2nd. ed.). Thousand Oaks, CA: SAGE

62. Miles, R.E., Snow, C.C., Fjeldstad, O.D., Miles, G., & Lettl, C. (2010). Designing organizations to meet 21st-century opportunities and challenges. Organizational Dynamics, 39(2), 93–103.

63. Nadin, S. & Cassell, C. (2004). Using data matrices. In C. Cassell & G. Symon (Eds.), Essential guide to qualitative methods in

organizational research. (pp. 271–287). London: SAGE. Retrieved June 23, 2014

64. Nesdill, D., Love, A., & Hunt, M. (2010). From subject selectors to college and interdisciplinary teams. Science & Technology Libraries, 29(4), 307–314.

65. Nowviskie, B. (2013). Skunks in the library: A path to production for scholarly R&D. Journal of Library Administration, 53(1), 53–66.

66. Nutefall, J.E., & Chadwell, F.A. (2012). Preparing for the 21st century: Academic library realignment. New Library World, 113(3/4), 162–173.

67. Ohio State University Libraries (2012). Overview of research commons in academic libraries: A white paper. Columbus, OH: Ohio State University Libraries. Retrieved April 30, 2014

68. Pacios, A.R. (2004). Strategic plans and long-range plans: Is there a difference? Library Management, 25(6/7), 259–269.

69. Posner, M. (2013). No half measures: Overcoming common challenges to doing digital humanities in the library. Journal of Library Administration, 53(1), 43–52.

70. Pugh, L. (1997a). Convergence in academic support services. London: British Library Research and Innovation Centre

71. Pugh, L. (1997b). Some theoretical bases of convergence. New Review of Academic Librarianship, 3(1), 49–66.

72. Pugh, L. (2005). The management of hybrid libraries. Library and Information Research, 29(92), 13–31. Retrieved April 30, 2014

73. Pugh, L. (2007). Change management in information services (2nd ed.). Abingdon: Ashgate

74. Quinlan, C., & McHarg, H. (2012). The emerging library: Structure, culture, and lessons learned from the dissolution of a combined libraries–IT organization. Journal of Library Administration, 52(2), 147–161.

75. Radom, R., Feltner-Reichert, M., & Stringer-Stanback, K. (2012). Organization of scholarly communication services: SPEC Kit 332. Washington, DC: Association of Research Libraries

76. RCUK (2013). RCUK policy on open access and supporting guidance. Swindon: Research Councils UK. Retrieved April 30, 2014

77. Renfro, P., & Neal, J.G. (2012). The integration of libraries and academic computing at Columbia: New opportunities for internal and external collaboration. Journal of Library Administration, 52(2), 162–171.

78. Schrader, A., Shiri, A., & Williamson, V. (2012). Assessment of the research learning needs of the University of Saskatchewan librarians: A case study. College & Research Libraries, 73(2), 147–163. Retrieved April 30, 2014, from http://crl.acrl.org/ content/ 73/2/147.full.pdf

79. Shore, E. (2012). Embracing hybridity: The merged organization, Alt/ac and higher education. Journal of Library Administration, 52(2), 189–202.

80. Stanford, N. (2013). Organization design: Engaging with change (2nd ed.). Abingdon: Routledge

81. Stanton, J.M., Kim, Y., Oakleaf, M., Lankes, R.D., Gandel, P., Cogburn, D., & Liddy, E.D. (2011). Education for eScience professionals: Job analysis, curriculum guidance, and program considerations. Journal of Education for Library and Information Science, 52(2), 79–94

82. Vandegrift, M., & Varner, S. (2013). Evolving in common: Creating mutually supportive relationships between libraries and the digital humanities. Journal of Library Administration, 53(1), 67–78.

83. Vaughan, K.T.L., Hayes, B.E., Lerner, R.C., McElfresh, K.R., Pavlech, L., Romito, D., Morris, E.N. (2013). Development of the research lifecycle model for library services. Journal of the Medical Library Association, 101(4), 310–314.

84. Vinopal, J., & McCormick, M. (2013). Supporting digital scholarship in research libraries: Scalability and sustainability. Journal of Library Administration, 53(1), 27–42. Retrieved June 23, 2014, from https://archive.nyu.edu/bitstream/2451/31698/2/Vinopal_McCormick_Final.pdf

85. Warren, V. (2011). Using workforce structures to enable staff retention and development: An academic library case study. New Library World, 112(1–2), 8–18.

86. Yakel, E., Conway, P., Hedstrom, M., & Wallace, D. (2011). Digital curation for digital natives. Journal of Education for Library and Information Science, 52(1), 23–31. Retrieved June 23, 2014

87. Young, H., & Lund, P. (2008). Reflections on a benchmarking survey of research support provided by 1994 Group libraries. SCONUL Focus, 43, 51–56. Retrieved April 30, 2014

Chapter 3

Research Libraries' New Role in Research Data Management, Current Trends and Visions in Denmark

Filip Kruse, Jesper Boserup Thestrup

The State and University Library, University Library Division, Aarhus, Denmark

ABSTRACT

The amount of research data is growing constantly, due to new technology with new potentials for collecting and analysing both digital data and research objects. This growth creates a demand for a coherent IT-infrastructure. Such an infrastructure must be able to provide facilities for storage, preservation and a more open access to data in order to fulfil the demands from the researchers themselves, the research councils and research foundations.

This paper presents the findings of a research project carried out under the auspices of DEFF (Danmarks Elektroniske Fag- og Forskningsbibliotek - Denmark's Electronic research Library) [i] to analyse how the Danish universities store, preserve and provide access to research data. It shows that they do not have a common IT-infrastructure for research data management. This paper describes the various paths chosen by individual universities and research institutions, and the background for their strategies of research data management. Among the main reasons for the uneven practices are the lack of a national policy in this field, the different scientific traditions and cultures and the differences in the use and organization of IT-services.

This development contains several perspectives that are of particular relevance to research libraries. As they already curate digital collections and are active in establishing web archives, the research libraries become involved in research and dissemination of knowledge in new ways. This paper gives examples of how The State and University Library's services facilitate research data management with special regard to digitization of research objects, storage, preservation and sharing of research data.

This paper concludes that the experience and skills of research libraries make the libraries important partners in a research data management infrastructure.

KEYWORDS

research data management; university; library; national cultural heritage

1. INTRODUCTION

The focus of this paper is on how national libraries and university or research libraries have a new and important part to play in research data management. First, in Section 2, we outline the concept and changing reality of research data and we make a brief presentation of the results of a DEFF research project (Danmarks Elektroniske Fag- og Forskningsbibliotek — Denmark's Electronic Research Library) that charted the field of how research data management is carried out in the universities of Denmark. The activities connected with research data management such as providing facilities for data reuse and data sharing change the concept and reality of research data. We discuss this in Section 3. As part of its obligation to preserve the national cultural heritage, The State and University Library is already providing an e-infrastructure for researchers' access to its radio, television and web archives. In Section 4 we show how the cooperation between researchers and library staff is a precondition for developing and providing the right facilities for access to, and use of, data.

2. THE DEFF PROJECT RESEARCH DATA MANAGEMENT IN DENMARK

2.1 Aim of the project

In 2013, a project group finished a project initiated and financed by DEFF with the publishing of the report *Forvaltning af forskningsdata I Danmark* (Research Data Management in Denmark) (**Thestrup et al., 2013**). The aim of this project was to gain knowledge on a subject hitherto unexplored in Denmark, namely how the Danish universities' work on research data management is organized and how it is carried out in practice, especially with regard to data storage, data preservation and data sharing or other forms of making data more openly accessible.[2] All Danish universities participated in the survey.

Closely connected to the aim of establishing an overview in the form of a catalogue of knowledge of the universities' current activities and future plans in this field, it was also important to supplement this with the most significant international experiences in this field. In order to complete the picture of the current situation in the field of research data management in Denmark, it was also necessary to explore the leading Danish research councils' and foundations' viewpoints towards increased data accessibility. The overall vision for the project was to increase the attention and the interest of decision makers in this field, on the political level as well as on the university level.

2.2 Methods employed

The study is explorative in its design and its focus is on the present state of research data management at the level of policy and strategy, and not on the researchers' specific activities.

The project group carrying out the survey consisted of participants from all the universities' libraries: The Royal Library,[3] Copenhagen Business School Library,[4] Aarhus University Library,[5] State and University Library,[6] DTU Library,[7] University of Southern Denmark Library,[8] Aalborg University Library[9] and Roskilde University Library.[10]

The survey was carried out as a series of personal interviews with top level managers from the universities, university libraries and university IT-departments. The interview topics[11] covered the individual university's plans and activities with regard to data storage, data preservation, data archives, enhancing the researchers' awareness of data preservation, visibility and accessibility, and whether the university's exposure of research data is considered to be a part of its branding strategy. The personal interview was chosen, as the data-collecting technique since it — in contrast to the ordinary questionnaire-based survey — permits dialogic interaction such as follow-up questions to the interviewee's answers. Regarding the research councils and foundations, a slightly altered questionnaire was issued based on the interview guide, and it was distributed by e-mail.

Our intention by the first three seemingly overlapping topics was to ensure that the interview would cover as wide a range as possible of the various activities undertaken by universities to secure research data and to make them accessible during and after the research process.[12] Likewise, we used the term 'research data' in the interview guide in a broad sense, as the aim was that the interviewees should give as much information as possible on their activities in this field. Data storage is essentially data backup. We know from this and other surveys in the field (**Sørensen et al., 2009**; **Pattenden-Fail et al., 2010**), that researchers' backup actions span from the use of memory sticks, over external hard drives and mail systems to the use of various cloud services. General characteristics of data storage or data backup are that the data are not meant to be accessed by others than the creator(s) and consequently they are not issued with metadata or other means of making them searchable. Further, changes will be made to the data during the research process and the first data collected will not necessarily be saved.

2.3 Concepts and definitions

Data preservation, in contrast to data backup, means securing permanent access to the original research data, and, as a rule, data from the finished research project. General characteristics of data-preservation actions are that the data are accessible to others for verification — such as datasets as part of scientific publications — or for sharing or collaboration within the scientific community. Consequently, the data must be organized and made searchable within institutional settings such as universities' repositories or data archives, or other data archives. Conditions for access to data and for reuse are formally regulated. Long-term preservation of the original data and protection of sensitive data are vital characteristics of data-preservation actions.

Data archives[13] are institutions for data preservation; they are mostly subject-specific, organized on an institutional, national or international level. National data archives are frequently older institutions than, e.g., institutional repositories.

3. MAIN RESULTS OF THE PROJECT

3.1 Data storage

Apart from a few exceptions, Danish universities, as a rule, provide researchers only with general access to the university computer drives, and not with special facilities for data storage. The researchers themselves decide upon which type of storage they want to use, provided that it is secure. The responsibility often lies with the university department. The reasons for this policy, as explained by the respondents, are that data storage, preservation and sharing are not considered to be as important as the publication of research results. Another reason is that considerable uncertainty exists regarding the definition of research data and how to store or preserve them securely. This is reflected in the absence of centralized responsibility in favour of a transfer to departments and individual researchers, in keeping with a sceptical attitude towards a uniform (national) policy and common technological solutions — a so-called 'one-size-fits-all' arrangement. The multitude of different data types and specific needs in relation to storage are also arguments in favour of this.

The research councils and foundations make no specific demands on data storage for the research projects they have funded but expect legal and academic standards to be followed.

3.2 Data preservation

This activity is, in general, considered too resource-demanding for the universities to undertake, and with a 10+ years' time scale, it is not regarded as a natural task for the universities. Two universities have storage facilities which can also be used for long-term data preservation. There are only a few examples of research data being preserved by universities. Again, the different data types and preservation needs, together with the lack of legal obligations, are arguments for this somewhat hesitant policy. Correspondingly, not much attention from the managerial level is given to the potentials of making research data accessible or shareable. The research councils and funds have no formulated policy for data preservation but regard it, to some extent, as the responsibility of the institutions.

3.3 Data archives

The universities favour a national solution consisting of several data archives as a supplement to, and an expansion of, the existing archives. One university favours international data archives based on its specific experiences in 'big science'. The arguments for the national solution are the qualitative and quantitative differences in needs, such as data types, academic traditions and cultures and size of datasets, and also research cooperation between universities, common principles for access to data and financial costs. A national solution also calls for cooperation between the existing data archives. A few universities are, at present, considering establishing their own archives but are awaiting results from more detailed analyses.

The research councils and foundation funds are, in principle, in favour of making data accessible or shareable as this will improve utilization of the resources. This issue must be deliberated in the context of a national research infrastructure, the existing data archives, a coherent definition of research data, standardization of metadata and legal matters in connection with data sharing.

3.4 Researchers' awareness and research data as university branding

The majority of universities is already carrying out campaigns or other activities — or is planning to do so — with the aim of increasing the researchers' awareness of the potentials of preservation and accessibility of research data. All the universities in this study emphasize the importance of publication of research results as part of their strategy, and apart from new ground-breaking data, publications are considered more important than research data, as elements of strategic branding.

3.5 Perspectives

A general result of the survey is that activities in the field of research data storage, preservation, archiving and sharing vary from university to university. This is probably due to differences in institutional and organizational structure, in size and in academic profile. Further, it appears that the universities do not seem to be very active in the field. How should we interpret this?

One way of seeing this is that the universities take a position best described as 'wait-and-see' on the issue of research data. But at the same time, they express concern in relation to the researchers' needs and do not want to exert pressure on them, e.g., in the form of mandatory rules for data preservation. Decentralized solutions are preferred, as they are seen as potentially more suitable to local needs and available resources. This consideration could also be seen reflected in the emphasis expressed on issues related to different needs, different data types, various sizes of datasets, different academic cultures, etc. The need for decentralized solutions can be seen in the cases we describe in Section 4. Researchers were involved in designing the software for handling the research objects, in order to meet their specific needs.

In addition, the launching of various local projects could be regarded as attempts to explore this mainly uncharted territory. A supplementary explanation could be that the universities are unsure of how they should handle the task, e.g., because of lack of experience in the matter.

Funding is a separate concern, and publication of research results is a prerequisite for financial support for research projects, while data preservation and sharing are not. At the same time, both universities and research councils and foundations emphasize — in our opinion rightly — the importance of a clear definition of research data, standardization of metadata, clarification of central legal matters in relation to data sharing and a clear division of labour in relation to the data archives.

A quote from the questionnaire of one of the Danish research funds:

"The fund believes that data archives should be national or international, not institutional. The recent fusions, divisions and closing down of e.g., university departments demonstrates that the individual university's infrastructure has a relative short life and is ill-suited for long-term preservation of research data."

In our opinion, the universities' position is clearly understandable, as they attempt to navigate in a space characterized by different needs in the various fields of research, absence of clear definition of the problem's nature, lack of financial incentives, lack of a national policy on the issue and perhaps also lack of experience in data preservation, use of metadata and sharing of research objects. The national and research libraries already have a stable organizational position and are currently involved in various projects here. A few examples to illustrate this point:[14] recording and preservation of research data in two universities (ORBIT, Danish Technical University, RUDAR, Roskilde University), software development (DATAVERSE, The Royal Library), archiving the Danish internet (Netarchive, The State and University Library) and archiving Danish radio broadcasts (LARM, The State and University Library). The libraries have skills and experience in generating metadata for research objects and datasets, in preservation and in facilitating access to and use of data. Therefore, the libraries are the researchers' natural collaborators in developing and implementing the systems necessary for the management of research data.

Based on the findings of the survey, the project group has submitted several recommendations, among them, that a national policy, as well as a policy for the universities for preservation and sharing of research data, is formulated and that the necessary e-infrastructures are established.

A study of European and international practices in research data management15shows some interesting different findings. Here, funders and publishers are found to be the main drivers for work on data management plans and thus also contribute to a national policy in the field. In Denmark, the research councils and foundations are not particularly active, but seem to be waiting for a national policy, rather than developing policies of their own. The Danish universities are in the process of implementing research data management activities, but both the universities and the research funders seem yet to be lagging behind compared to the European and international development, but e.g., DeIC's activities indicate a growing awareness of the issue.

3.6 Same play, new actors — an update

Since the DEFF project was completed, several different institutions in Denmark have been active in developing infrastructures for research data management.

Various activities have been carried out at the individual university level in order to ensure internal processes regarding management of research data. At the national level, DeIC (the Danish E-Infrastructure Cooperation)[16] and DEFF are working to evaluate different services involved in data management from application for grants to preservation and sharing of data. DeIC was founded in 2012, with the primary task to "support Denmark as an e-Science nation through delivery of e-infrastructures." DeIC was established as a merger between Forskningsnettet (the Danish Research Network) and the Danish Center for Scientific Computing (DCSC).

DeIC has formulated six strategic goals. One of them is to "Coordinate solutions concerning Data Management and large datasets"[17]:

*"The amount of research data rises continuously as do the demands for long term preservation. The possibility to reuse and share data across research groups is also a criterion. The institutions' interest in and need for solutions are urgent. Stakeholders: the universities, the research libraries and other research institutions." (**DeIC, 2012**)*

One of the projects in which DeIC is involved is FIF (Fælles Infrastruktur for Forskningsdata — Common Infrastructure for Research Data). The aim of this project is to develop a Danish infrastructure for data management (**Christensen-Dalsgaard, 2013**). The project, which is based on a grant from DEFF, involves resources from DeIC as well. Since DEFF represents libraries and DeIC represents the universities, the project demonstrates that research libraries are regarded as natural partners by the universities, regarding management of research data.

Parallel to this, DeIC has initiated a process to ensure that a national strategy for data management is formulated. In order to do that, DeIC has contacted the Danish universities, the Danish research councils, The State and University Library, The Royal Library, DDA (the Danish Data Archive), DEFF and a research infrastructure project called DigHumLab (Digital Humanities Laboratory[18]). DeIC wants to formulate a policy which can be approved by the relevant institutions and by the Ministry of Science, Innovation and Higher Education19 (**DeIC, 2013**).

Later we shall show that some of the services provided for researchers by The State and University Library in connection with its obligation to the preservation of the national cultural heritage, already facilitate the reuse and sharing of research data. Facilities for sharing research data or other forms of improving data accessibility are an important part of the requirements of data management systems (see, e.g., **Higgins, 2012**). Further, we shall discuss future perspectives generated by these services for the role of research libraries. But before we turn our attention to these matters, we shall consider how the concept and reality of research data has changed, and discuss the possible implications of this development.[20]

4. FROM RESEARCH DATA TO RESEARCH OBJECTS AND VICE VERSA

4.1 Research data

In order to provide a conceptual framework for further analysis it is necessary briefly to consider the concepts 'research results', 'research objects' and 'research data'. Several initial definitions are provided by the Data Information Specialist Committee — UK:

"...research data, that which is collected, observed, or created, for purposes of analysing to produce original research results. This differs from what is commonly called research outputs, which are the peer reviewed, published papers/articles/books/presentations that are produced as a result of data analysis. Research data may be created in tabular, statistical, numeric, geospatial, image, multimedia or other formats." (http://www.disc-uk.org/qanda.html)

Research results are outputs of the research process, which has its focus on a specific research object. Research data are thus always data 'on' a specific research object, for instance, Stone Age tools, solar flares or political extremism in Europe. Research data can be seen as records of the research activity, i.e., that which is created by the activity and preserved for use or reference in the future. Consequently, the records must be authentic, reliable, usable, complete and unaltered (**Higgins, 2012**, p. 20). Research data can be primary or secondary.

"Primary data are data that are collected for the specific research problem at hand, using procedures that fit the research problem best. On every occasion that primary data are collected, new data are added to the existing store of social knowledge. Increasingly, this material

*created by other researchers is made available for reuse by the general research community; it is then called secondary data." (**Hox & Boeije, 2005**, p. 593)*

This distinction is based on difference in the purpose of the data collection and consequently also on its relation to the research process: Whereas primary data are collected with the purpose to contribute to solving a specific research problem, secondary data are collected with different research purposes than the one for which they are initially used.[21] The researcher using secondary data "by an act of *abstraction* uses questions originally employed to indicate one entity to illuminate other aspects that a former analyst did not have in mind at all." (**Hyman, 1972**, p. 37)

Primary data can be collected through methods such as experiments, clinical tests, qualitative interviews and surveys, depending on the specific research project and the area of science to which it belongs. Social science researchers often either produce their own data through methods such as surveys, interviews and field studies, or they use data from statistical records. Science data are also produced by the researchers, often by observations, experiments and computer models, or they are drawn from data archives, e.g., researchers in astronomy and genomics have for some time shared their data in common archives (**Borgman, 2012**). Humanities data "most often are drawn from records of human culture, whether archival materials, published documents, or artefacts" (op. cit. p. 1061).

Primary and secondary data can be quantitative or qualitative. Secondary data used as basis for research are often quantitative and can be obtained from sources such as national statistical databases and government archives, and qualitative data can also similarly be found in some national data archives such as The Danish Data Archive.[22] When data are made accessible by depositing or sharing they can be reused for other research purposes, if primary data are reused they change into secondary. Below we elaborate further on the implications for the concepts of research object and research data.

4.2 Access to research data

A large international survey of researchers' practices and perceptions regarding data sharing finds that the majority of researchers from all subject disciplines have a positive attitude to sharing their own data and uses others' (**Tenopir et al., 2011**). Only a minority, however, does so in the real world. Among the important conditions for agreement to share is a guarantee for authorship credit in the form of proper citation, offers of collaboration or of financial contributions to research. Among the reasons given for not making data available to others are lack of time, lack of funding, lack of places to put data and lack of standards. Still, at the same time, lack of access to other researchers' data is regarded as a barrier to scientific progress.[23]

A survey of researchers' practice at Aarhus University shows that they favour a more open access to research data and at the same time clearly express a need for better facilities for data storage and preservation (**Sørensen et al., 2009**). Similar results — in relation to issues of access — are found in an interview-based study of researchers at Glasgow University (**Pattenden-Fail et al., 2010**).

In a short summing up of the present circumstances, two trends are identified: at the same time as subject-specific data archives flourish, a lot of data lie on personal hard drives or are saved (or forgotten) elsewhere (**Nelson, 2009**). This is due to several factors: 'where', the lack of infrastructure, i.e., databases suitable for the various fields of science, 'how', basically the question of data standards, formats and metadata and 'which', raw data or quality controlled data?

An interpretation of these findings could be that issues of copyright in the sense of scientific recognition of authorship of data plays an important role, but we should also note the above-mentioned 'where', which points to insufficient data infrastructure and the 'how', which is the lack of common standards, both making data sharing cumbersome and time consuming for the researchers.[24] Whereas the issue of forms of recognition of authorship to research data is a matter that could be regarded as primarily belonging to the scientific community, the issues of infrastructure and data standards are rather a matter for information professionals in cooperation with the researchers.

But why share research data? The arguments for giving access to research data and thus facilitating sharing, use and reuse of data can be summarized thus, drawing on **Borgman (2012)**:

1) Reproducing and verifying research results. This is, to some extent, an ideal standard, as not all experiments and field studies can be replicated. In the sense of verification, however, it is stressed that scientific work is subject to tests of validity and reliability by the scientific community.

2) Serving the public interest. This can be seen both as the 'tax-payer argument' and as an argument for democratic transparency.

3) Enabling new questions to be asked for research data and results. This is related to 1, but not identical to it, as it emphasizes the possibility of reuse of data in the form of the combination of different data. It is also related to 2.

4) Advancing new research. The underlying argument is that data sharing can advance scholarship. This will, in turn, increase returns from investment in research. This is also related to 2.

Of vital importance for the access to and sharing of research data is that they can be discovered and retrieved. The condition for this is that they are described by means of metadata. As the aim of this study is not an in-depth discussion of metadata and their proper use, we limit ourselves to Higgins' typology of metadata (**Higgins, 2012**, p. 38):

- Descriptive metadata — ensures identification, retrieval, classification, links to related resources;
- Technical metadata — records file formats, software or hardware;

- Administrative metadata — information on acquisition, accession and issues of intellectual property rights;
- Use metadata — manages access, use statistics;
- Preservation metadata — documents preservation, migration.

4.3 Research data and research objects — the distinction revisited

It follows from the above reflections on data and objects that the original distinction between primary and secondary data cannot be maintained without modifications. If primary data are shared and reused by other researchers for other research aims, they must be regarded as secondary data. However, the aim could also be identical, but be for the purpose of the verification of earlier results (cf. 1.). The possibility of asking new questions to old data or of a reuse together with new (primary) data (cf. 3.) also blurs the original distinction between primary and secondary data.

Research objects, such as documents and recordings can be stored in digital archives and made searchable by issuing them with metadata. As we shall see later, this is the case for The State and University Library's digital archives such as LARM (LARM Audio Research Archive),[25] The Netarchive[26] and Mediestream.[27] In the case of LARM, users can already share annotations to sources (radio broadcasts) and in the case of Mediestream, users' enrichment of the available metadata through crowd sourcing is a future vision. We will later show that these cases illustrate that research objects can be issued with potential research data and, in turn, made accessible and shareable. This makes them at the same time research objects — and research data. It expands the scope of access to research data and brings the research libraries' and national archives' role into focus. We are also faced with the need for a wider understanding of the new nature of research data, facilitated by digital preservation. Indirectly, this also strengthens the argument for the use of the term 'research data' in the wide sense in the interviews of the DEFF-project.

5. E-INFRASTRUCTURE FOR THE DANISH NATIONAL CULTURAL HERITAGE — TARGETED AT RESEARCHERS

Below we explore the ways in which the various digital preservation projects carried out by The State and University Library contribute to increasing the accessibility of research data. We concentrate on the services LARM, Netarchive, Mediestream, and DigHumLab. We will use the insight gathered from the DEFF project to explore the practice of data management with such question as: Are data preserved, curated, or 'only' stored? Which facilities are available?

Of special importance are the options for the users' adding of data to the digital objects, changing or enriching the metadata and sharing of data. These services provide access to original research objects (audio and video recordings, websites etc.) that are stored and preserved by The State and University Library. When they, for instance, are used and manipulated by researchers, enriched with additional data and eventually made accessible to other researchers, they change their nature and become both research objects and research data. These opportunities to change and add data requires that the storage system can handle the original data, as well as metadata about the changes and the manipulated data.

In our elaboration of the cases, we focus on the factual role of the library and on the types of institutions also involved in the work. We believe that it is necessary to consider whether a project is solely a project under the auspices of libraries, archives and museums (the LAM sector) or whether it also has participation from the research sector. Finally, we use our description of the services to demonstrate how a national and research library can play a vital role in research by establishing infrastructures suited to the researchers' needs.

The cases we present below illustrate how the library — here the State and University Library — collect the data while the researchers cooperate with the library in the development of software for the use of the data.

5.1 Larm.Dk

LARM is a project active from 2010 to 2013.[28] In total, 11 different institutions are involved, among them 5 universities,[29] The Kolding School of Design, The Royal School of Library and Information Science, DeIC, the Danish national broadcast service (DR), The Museum of Media and The State and University Library. LARM has established a database with radio programs from 1925 to date. The purpose of LARM is to enable researchers to use radio programs as a source for humanistic research.

The State and University Library has been involved in LARM in different ways. One way has been as a source of data for the service, especially with scanned documents which has undergone OCR, in order to enrich the metadata of files originating from DR. By the end of 2013, the database will contain more than 1,000,000 hours of radio airtime. LARM has developed a platform to present the objects and data from the database, and to enable the researchers to annotate the metadata of a given program and to collaborate using each other's annotations. LARM is an example of research infrastructure, providing access to data and objects via metadata, with facilities for annotation, sharing and collaboration. Access to the database itself is limited to researchers. The annotations are shared as text under a Creative Commons License.[30] The archive itself is not intended as a facility for long-term preservation. However, the data are preserved by The State and University Library.

The platform which makes this possible is called CHAOS (Cultural Heritage Archive Open System[31]) and was developed by DR, The State and University Library and the Danish Research Network.[32] The researchers access the radio programs via LARM.fm [33] when they are logged into the system; they can search the files, make their annotations and share these annotations. During this process the original objects are supplied with primary research data (or metadata), which, in turn, can be reused as secondary data by other researchers (**Andersen, 2012a**; **Nordicom-Information, 2012**).

LARM is an example demonstrating that close cooperation is needed between the LAM sector and the researchers in order to develop a system for access to digital research objects, to supply their metadata and share the enriched objects. The LAM sector has made both data and knowhow available regarding how to store large amounts of data and make them searchable. The LAM sector provided this project with knowhow regarding software programming.

5.2 The Netarchive — the Danish Web archive

Since 2005, The Royal Library and The State and University Library have harvested the Danish part of the dynamic internet.[34] From 2005 on, this has been required by law. The harvesting is done by using several mandatory strategies:

1) Broad crawls: 4 times each year all websites from the domain.dk and Danish websites registered under other domains like .com, .org etc. are captured.

2) Selective crawls: 80–100 specially selected sites are harvested. These are harvested with different intervals, some monthly, some up to six times a day.

3) Event crawls: If a given event is rated as being of importance for the Danish society, and the pages are expected to disappear after the event, the archive harvests the related websites.

The Netarchive illustrates that institutions in the LAM sector are active with regard to important prerequisites for research infrastructures, in this case storage and long-term preservation of research objects which can be used by infrastructure services.

In 2012, 992,822 websites were harvested. This, of course, generated a huge amount of data. 88,542 GB are now stored in the archive. The material is issued with metadata and long-term preservation. The web pages thus harvested are archived as objects, which when used by researchers, are transformed into research data. Collaboration in the form of data sharing, as in LARM, is not yet possible. Only researchers can have online access to the Netarchive; if you are not a researcher but have a scientific purpose you can access it from the premises of the involved institutions. These limitations are necessary in order to avoid problems with copyright and to ensure the necessary protection of personal data (**Andersen, 2012b**; **Schostag & Fønss-Jørgensen, 2012**).

The software which gives access to the data is constantly being developed. At first, each researcher literally had to have personal assistance in order to search and use the archive. Then the Netarchive started to develop software, based on the Internet Archive Wayback Machine,[35] in order to improve online access. The archive is now involved in a project called the Digital Humanities Laboratory (DigHumLab),[36] whose goals, among others, are to improve the search facilities and to create new ways to analyze the materials in the archive. Below we shall go more into details about DigHumLab (**Schostag & Fønss-Jørgensen, 2012**).

The Netarchive illustrates that institutions in the LAM sector are actively working in fields that contain the key elements of infrastructures for the management of research data such as storage and long-term preservation of research objects and providing access for users.

5.3 Mediestream

Mediestream[37] is a project initiated in 2010 by The State and University Library. The project started as an attempt to digitize part of the national cultural heritage, which The State and University Library has in its collections. The project involves a newspaper collection, collections of TV and radio broadcasts and several other collections. Mediestream now contains TV programs that would take more than 87 years to watch. The newspaper collection contains 32 million pages from Danish newspapers from 1666 to today. The amount of data involved is massive, both in number of files involved and overall size. The newspaper collection will be digitized during the next couple of years.

The objects in Mediestream will be digitized material and material which was digitally born or later digitized as part of the archiving process. For example, the newspapers will be digitized using microfilms as the source. The TV programs will partly be analogue material digitized from VHS- based collections dating back from 1987, and partly be programs digitally harvested. Currently, Mediesteam offers access to Danish radio and TV programs broadcasted from 2006 to today, TV commercials from 1988 to 2005 and commercials shown in Danish cinemas from 1954–1995. The data will be long-term preserved.

For copyright reasons, the collection will be publicly accessible only on the premises of the Library, but it is online accessible for researchers and students (**Elstrøm & Jensen, 2012**; **Kirring & Andersen, 2008**; **Williams, 2012**).

In order to facilitate collaboration, The State and University Library would prefer it to be possible for researchers and students to enrich the metadata of the objects and share these annotations in both the Netarchive and Mediestream. This would transform the data for archival data into a dynamic database of research objects. Currently it is regarded as a technological problem, as a problem of quality with regard to the quality of the enriched metadata and as a copyright problem. In the future, LARM.fm could offer a solution to these.

Mediestream shows how national libraries can create collections of digital objects, which successfully can be made available for researchers and for the public as well. In this respect, Mediestream is primarily a facility for storage and preservation of research objects which, in turn, can provide content for research infrastructure services.

The three cases above illustrate the work processes involved when large collections of digital materials are made accessible. In order to facilitate searching and curating the data, librarians are needed to perform the traditional task of adding metadata. It is not just a question of adding them to the individual file or object, but also a question of combining different collections in order to create appropriate metadata. In other words, one has to use the knowledge of the librarians of the different collections libraries. **Lauersen, Christiansen, and Olsen (2012)**, present how different sources are combined in order to enrich the metadata of files in LARM. In this case, it was possible to digitize printed material in order to create metadata so that LARM could present information on the programs broadcasted from as far back as 1925. The authors present a workflow which could make it possible to ensure that different collections are used in combination in order to produce metadata of as high a quality as possible.

5.4 Digital Humanities Laboratory (DigHumLab)

DigHumLab is a project with the aim to contribute to skills development, internationalization and innovation through a national focus on Digital Humanities in research, education and knowledge transfer. The aim is also to develop new methods and tools for analysis of data in the digital collections (**Nordicom-Information, 2012**). [38]

The project is carried out in cooperation between Aalborg University (AAU), Aarhus University (AU), The University of Copenhagen (KU), The University of Southern Denmark (SDU), the Royal Library and The State and University Library.[39] Besides these, the project also has international partners.[40]

The project addresses three different fields: Language Based Materials and Tools, [41] Media Tools[42] and finally Interaction and Design Studies.[43] We do not go into depth with each of these areas, but focus on Media Tools. This provides tools for two areas — each with a work package — which will enable the researchers to analyze web pages; also it will facilitate new ways of working with sound and media. Both Media Tools work packages involve The State and University Library. The first involves the Netarchive and the second involves the collections of films, TV and Radio that The State and University Library stores as part of its obligation as a national cultural heritage library (**Nordicom-Information, 2012**).

In order to ensure that the researchers involved in DigHumLab can access and use the data, The State and University Library is cooperating closely with the relevant researchers to develop the necessary tools. DigHumLab can be seen as an example of how the massive amounts of objects and data in the digital collections of national libraries best can be used by the researchers if they are involved in the development of tools for access and analysis.

5.5 Summary: access to research data and options for sharing and collaboration

The projects presented in this paper are summarized in Table 1.

Table 1. Overview of the projects.

Service	LARM	The Netarchive	Mediestream	DigHumLab
Purpose	To make Radio programs available as a source for research	To preserve Danish websites	To digitize and present cultural heritage online	To improve access to digital collections and new methods for analyzing the data
Content	Radio programs and other material from 1925 to today	Harvested websites	Different collections of The State and University Library	A wide range of online information, depending on the different projects
Participating institutions	Universities, institutions from the LAM sector, DR	The Royal Library and The State and University Library	The State and University Library	Universities, institutions from the LAM sector, DR
Data sources	DR, The State and University Library	The Netarchive	The collections of the participating institutions	Depending on the projects, among them the Netarchive and Mediestream
Access	Researchers — online	Researchers — online	Researchers — online and on the premises for others	Researchers — online
Long term storage	Yes	Yes	Yes	In the archives — yes
Options for users to annotate and collaborate	Yes	No	No	Part of the project is to ensure requirements for researchers' collaboration

The cases presented above demonstrate that a national library like The State and University Library, with a wide range of tasks, must be regarded as a natural part of a system which handles research data. In order to fulfil its tasks, the library has obtained experience, skills and knowledge regarding preservation of massive amounts of data, both in number of files and in size. These competences are created in cooperation between the LAM sector, researchers from the universities and DR, and are offered to other institutions outside the LAM sector via Digitalbevaring.dk. 44 The tradition of cooperation is continued in the FIF project mentioned earlier, in which DeIC and DEFF are evaluating systems and software to handle research data.

6. CONCLUSION AND PERSPECTIVES

The Danish universities clearly have a positive position on the issue of storage, preservation and sharing of research data. What for most universities prevents this from being translated into practice, is the lack of a national policy, diverse needs from the researchers' part depending on culture and traditions in the different fields of research, differences in IT-infrastructures and a general lack of financial incentives.

To some extent, the reuse of data as a consequence of data sharing blurs the classic distinction between primary and secondary data. In the case of digitally archived research objects such as broadcasts and websites, bringing together the examples of The State and University Library's services, the researchers' adding of metadata and annotations to the objects and making them shareable, adds new aspects to data sharing. This reflects a more collaborative research culture also inherent in the concepts of 'data deluge' and 'fourth paradigm for science' (the analysis of massive datasets) and also emphasizes the need for professional handling of copyright issues.

Providing storage and preservation facilities for researchers on an institutional and national level in the form of e-infrastructure(s) is the logical answer to the demand for a more open access to research data and research objects such as items of national cultural heritage. A viable e-infrastructure not only implies technology-based services for the description of data by metadata to ensure proper identification, facilitate retrieval and reuse, but also human expertise. The fields such as those of open access, copyright, metadata and archiving are evident. Less evident, but equally important, is the subject-specific knowledge of the different data collections. This is a prerequisite not only for the adequate description of data, but also for understanding the researchers' needs and the transformation of these into adequate technical solutions. This is best carried out in collaboration between the library staff and the researchers, as demonstrated by the cases from The State and University Library.

In general, the universities, the national libraries and national archives are responding to the challenge to improve the conditions for data management, data sharing and data preservation. DEFF, DeIC and several research libraries are involved in a project to evaluate different systems to handle research data. DeIC has been in contact with universities, national archives and research libraries in order to formulate a national policy regarding research data management. The universities themselves are formulating their own internal policies regarding research data. In all, one must expect that in a short time, the universities and the LAM sector will establish a coherent national system for the handling of research data.

NOTES

[1]http://www.deff.dk/english/. Retrieved August 12, 2013.

[2]These topics are only parts of a data management plan, Donelly's checklist overview contains 10 items (**Donelly, 2012**, pp. 93–94).

[3]http://www.kb.dk/en/index.html. Retrieved August 12, 2013.

[4]http://www.cbs.dk/en/library. Retrieved August 12, 2013.

[5]http://library.au.dk/en/. Retrieved August 12, 2013.

[6]http://en.statsbiblioteket.dk/. Retrieved August 12, 2013.

[7]http://www.dtic.dtu.dk/English.aspx. Retrieved August 12, 2013.

[8]http://www.sdu.dk/en/Bibliotek. Retrieved August 12, 2013.

[9]http://www.en.aub.aau.dk/. Retrieved August 12, 2013.

[10]http://rub.ruc.dk/en/. Retrieved August 12, 2013.

[11]For a full list of the interview questions and sub-questions see **Thestrup et al. (2013)**, Appendix 8.

[12]This does not include data that cannot be made public for data-protection reasons or data from private companies' research.

[13]See **Doorn and Tjalsma (2007)** for an overview of the development and recent trends of data archiving.

[14]http://rudar.ruc.dk, Retrieved August 12, 2013. https://data.kb. dk/dvn/, Retrieved August 12, 2013. http://www.larm-archive.org, Retrieved August 12, 2013.http://netarkivet.dk, Retrieved August 12, 2013.

[15]http://www.sim4rdm.eu/docs/project-outputs/sim4rdm-landscape-report. Retrieved November 1, 2013.

[16]http://www.deic.dk/node/110?language=en. Retrieved August 12, 2013.

[17]http://www.deic.dk/strategi. Retrieved August 12, 2013.

[18]http://dighumlab.dk/about/. Retrieved August 12, 2013.

[19]http://fivu.dk/en. Retrieved August 12, 2013.

[20]Part of this discussion is related to what is known as the "Fourth Paradigm of Science" (**Hey, Tansley, & Tolle, 2009**).

[21]We do not follow the axiom that primary data *per se* must be collected by the researcher(s) exclusively (e.g., Data Information Specialists Committee-UK,http://www.disc-uk.org/qanda.html) as this, in our opinion, tends to lead to a categorization based on copyright issues rather than on the relation to the research process (see also, **Hox & Boeije, 2005**).

[22]For an analysis of the emergence of national data archives see, e.g., **Doorn and Tjalsma, 2007**.

[23]We must remember that prestigious scientific journals such as American Economic Review, Nature and Science, for several years have required authors of articles to submit digital copies of documentation such as data, codes and protocols to the journal archives.

[24]This is also noted in the literature on the 'Fourth Paradigm' see, e.g., **Hey et al. (2009)**.

[25]http://www.larm-archive.org/about-larm/. Retrieved August 12, 2013.

[26]http://netarkivet.dk/in-english/. Retrieved August 12, 2013.

[27]http://www.statsbiblioteket.dk/mediestream/. Retrieved August 12, 2013.

[28]http://www.larm-archive.org/about-larm/. Retrieved August 12, 2013.

[29]The University of Copenhagen, Roskilde University, The University of Southern Denmark, Aalborg University, Aarhus University, http://www.larm-archive.org/about-larm/. Retrieved August 12, 2013.

[30]http://creativecommons.org/. Retrieved August 12, 2013.

[31]http://www.chaos-community.org/en/index.html. Retrieved August 12, 2013.

[32]The Danish Research Network is today part of DeIC.

[33]http://www.larm-archive.org/about-larm/om-larm/om-larm-fm/. Retrieved August 12, 2013.

[34]http://netarkivet.dk/in-english/. Retrieved August 12, 2013.

[35]http://archive.org/web/web.php. Retrieved August 12, 2013.

[36]http://dighumlab.dk/. Retrieved August 12, 2013.

[37]http://www.statsbiblioteket.dk/mediestream/. Retrieved August 12, 2013.

[38]http://dighumlab.dk/about/mission-and-partners/. Retrieved August 12, 2013.

[39]http://dighumlab.dk/about/, Retrieved August 12, 2013 and http://dighumlab.dk/about/mission-and-partners/, Retrieved August 12, 2013.

[40]CLARIN (http://www.clarin.eu/), Retrieved August 12, 2013; DARIAH (http://www.dariah.eu/), Retrieved August 12, 2013; ESFRI (http://ec.europa.eu/research/infrastructures/index_en.cfm?pg=esfri), Retrieved August 12, 2013; TELEARC (http://www.noe-kaleidoscope.org/telearc/about/), Retrieved August 12, 2013 and IIPC (http://netpreserve.org/about-us/mission-goals), Retrieved August 12, 2013.

[41]http://dighumlab.dk/research-themes/language-based-materials-and-tools/. Retrieved August 12, 2013.

[42]http://dighumlab.dk/research-themes/media-tools/. Retrieved August 12, 2013.

[43]http://dighumlab.dk/research-themes/interaction-and-design-studies/. Retrieved August 12, 2013.

[44]http://digitalbevaring.dk/. Only in Danish. Retrieved August 12, 2013.

REFERENCES

1. Andersen, B. (2012a). Being a national library in a research infrastructure landscape. Microform & Digitization Review, 41(3–4), 175–179.

2. Andersen, B. (2012b). statistik_bja_2012. Unpublished manuscript

3. Borgman, C.L. (2012). The conundrum of sharing research data. Journal of the American Society for Information Science and Technology, 63(6), 1059. doi:10.1002/asi.22634. Retrieved November 19, 201

4. Christensen-Dalsgaard, B. (2013). FIF – fælles infrastruktur for forskningsdata. Unpublished manuscript

5. DeIC. (2012). Strategic goal: Coordinate solutions concerning data management and large data amounts. Lyngby: Danish e-Infrastructure Cooperation. Retrieved November 19, 2013

6. DeIC. (2013). Diskussionsoplæg – National strategi for datamanagement. Lyngby: Danish e-Infrastructure Cooperation. Retrieved November 19, 2013, from http://www.deic. dk/ DMdiskussionsopl%C3%A6g

7. Donelly, M. (2012). Data management plans and planning. In G. Pryor (Ed.), Managing research data (pp. 83–104). London: Planet Publications

8. Doorn, P., & Tjalsma, H. (2007). Introduction: Archiving research data. Archival Science, 7(1), 1–20

9. Elstrøm, G.V., & Jensen, T.S. (2012). Planning for mass digitisation of newspapers: A castle, a shed or something in between? Microform & Digitization Review, 41(3–4), 129–139. doi:10.1515/mir-2012-0021

10. Hey, A.J.G., Tansley, S., & Tolle, K.M. (2009). The fourth paradigm: Data-intensive scientific discovery. Redmond, Wash.: Microsoft Research. Retrieved November 19, 2013, from http:// research.microsoft.com/en-us/collaboration/fourthparadigm/ 4th_ paradigm_book_complete_lr.pdf

11. Higgins, S. (2012). The lifecycle of data management. In G. Pryor (Ed.), Managing research data (pp. 17–46). London: Planet Publications

12. Hox, J.J., & Boeije, H.R. (2005). Data collection, primary vs. secondary. In K. Kempf-Leonard (Ed.), Encyclopedia of social

measurement (pp. 593–599). New York: Elsevier. Retrieved November 19, 2013, from http://igitur-archive.library. uu.nl/ fss/2007-1113-200953/hox_05_ data%20collection, primary% 20versus%20secondary.pdf

13. Hyman, H.H. (1972). Secondary analysis of sample surveys: Principles, procedures, and potentialities. New York: John Wiley

14. Kirring, E., & Andersen, B. (2008). Dansk radio og TV på statsbiblioteket. DF Revy, 31(3), 4–7.

15. Lauersen, D., Christiansen, K.F., & Olsen, L.L. (2012). Management of metadata for digital heritage collections. Microform & Digitization Review, 41(3–4), 151–158.

16. Nelson, B. (2009). Data sharing: Empty archives. Nature — LA English, 461(7261), 160.

17. Nordicom-Information. (2012). Aktuella forskningsprojekt. Nordicom-Information, 34(3–4), 105–136

18. Pattenden-Fail, J., Sørensen, A.B., Kruse, F., Thøgersen, J., Molloy, L., & Ballaux, B. (2010). Report on academic research practices. London: Planet Publications. Retrieved November 26, from http://www.planets-project.eu/docs/reports/ Reportonacademic-researchpractices.pdf

19. Schostag, S., & Fønss-Jørgensen, E. (2012). Webarchiving: Legal deposit of internet in Denmark — A curatorial perspective. Microform & Digitization Review, 41(3–4), 110–120.

20. Sørensen, A.B., Kruse, F., Thøgersen, J., Molloy, L., Pattenden-Fail, J., & Ballaux, B. (2009). Report based on DT/7 questionnaire. London: Planet Publications. Retrieved November 26

21. Tenopir, C., Allard, S., Douglass, K., Aydinoglu, A.U., Wu, L., Read, E., et al. (2011). Data sharing by scientists: Practices and perceptions. PloS One, 6(6), e21101. doi:10.1371/journal.pone.0021101. Retrieved November 19, 2013, from http://www.plosone. org/ article/info%3Adoi%2F10.1371%2Fjournal.pone.0021101

22. Thestrup, J.B., Kruse, F., Nondal, L., Dorch, B.F., Andersen, M., Blaabjerg, N.J., et al. (2013). Forvaltning af forskningsdata i danmark, DEFF. Retrieved November 19, 2013

23. Williams, K. (2012). Mediestream: The trials, tribulations and triumphs of making a digital collection available online. Microform & Digitization Review, 41(3–4), 171–174.

Chapter 4

Scholarship 2.0: Analyzing scholars' use of Web 2.0 tools in research and teaching activity

Licia Calvi[1], Maria Cassella[2]

[1]NHTV Breda University of Applied Sciences

[2]University of Turin

ABSTRACT

Over the past 15 years the Web has transformed the ways in which we search for information and use it. In more recent years, we have seen the emergence of a new array of innovative tools that collectively go under the name of 'Web 2.0', in which the information user is also increasingly an information producer (i.e., prosumer), by sharing or creating content.

The success of Web 2.0 tools for personal use is only partially replicated in the professional sphere and, particularly, in the academic environment in relation with research and teaching.

To date, very few studies have explored the level of adoption of Web 2.0 among academic researchers in their research and teaching activity. It is not known in what way how and how much Web 2.0 is currently used within research communities, and we are not aware of the drivers and the drawbacks of the use of Web 2.0 tools in academia, where the majority of people is focused either on research or on teaching activities.

To analyse these issues, i.e. the combined adoption of Web 2.0 tools in teaching and research, the authors carried out a survey among teaching and researching staff of the University of Breda in The Netherlands. This country was chosen mainly because it is on the cutting edge as far as innovation is concerned. An important driver in choosing the Breda University's academic community was the fact that one of the two authors of this survey works as senior researcher at this university.

The purpose of our survey was to explore the level of adoption of Web 2.0 tools among the academic communities. We were interested in investigating how they were using these tools in the creation of scientific knowledge both in their research and teaching activity. We were also interested in analysing differences in the level of adoption of Web 2.0 tools with regard to researchers' position, age, gender, and research field.

Finally, in our study we explored the issue of peer reviewing in the Web 2.0 setting. In particular, we investigated whether social peer review is regarded by researchers as a viable alternative to the current closed peer review system (single-blind or double blind).

We approached about 60 staff members, but only 12 faculty members completed the survey fully. This means that our results can only be regarded as exploratory, but we still believe that they represent a complementary perspective with respect to previous studies.

KEYWORDS

research activity; social tools; teaching; user behaviour; Web 2.0

1. INTRODUCTION

Seven years after its birth, the Web 2.0 has become a consolidated virtual "reality". Due to their ease of use, social media have become pervasive and extremely popular among web users. This social phenomenon is mainly fueled by personal needs and motivations, but Web 2.0 tools are now also beginning to devolve their communicative power in people's professional lives.

In the academic context, it is extremely interesting to explore how social tools support science research and teaching and how the adoption of these tools is affecting the traditional four phases of the scholarly communication system: creation, certification, dissemination and preservation. The broad adoption of Web 2.0 technologies by research communities is also a big challenge both for publishers and libraries. They may feel threatened in their respective roles by the disintermediation brought along with the adoption of these technologies. As a consequence they should find strategies to cope with this phenomenon. Particularly, the main challenge for libraries is learning to use social tools not only to develop new services in the environment of social media and to "connect with researchers, but also to enhance the research process" (**McMahon et al., 2012**).

In this paper we will explore how the research communities are using Web 2.0 tools in their research and teaching activities, whether there are any relevant disciplinary or generational differences in their use, what tools they adopt more willingly, what contents they are trying to convey through the use of social tools, what benefits they derive from their use, and what the main obstacles are to their adoption. We will analyze the results of a survey carried out in April 2012 among the university lecturers of two departments within the Academy of Digital Entertainment, i.e., the Media and the Game departments of a University of Applied Sciences in The Netherlands, the University of Breda.

1.1 Scholarly Communication in the Web

Since the beginning of the 1990s, Internet has deeply innovated the way in which scientists do research. The innovative tools of communication and network sharing (such as e-mail, discussion forums, newsletters and, most recently, digital repositories — e.g., the physics preprint ArXiv or the biologists' database GenBank) and *collaboratories* (**Finholt, 2003**) have strongly reconfigured the different phases of scholarly communication (i.e., creation, certification, dissemination, preservation and research evaluation). The Web as a platform acts as a glue for the scholarly communication phases, reduces barriers to entry, speeds up scientific communication and collaboration, fosters the interdisciplinarity and cross-fertilization of science, and increases the democratization of knowledge by offering new tools and models for the dissemination of science (i.e., the open access paradigm which advocates for an open and seamless dissemination of knowledge).

The Web 2.0 has empowered all the above-mentioned innovations. Not all scholarly communication phases and functions are affected in the same way by the Web revolution. The two phases which show major changes are the creation of scientific knowledge (**Bukvova, 2010**; **Ponte & Simon, 2011**) and research dissemination. Preservation, certification and evaluation are less deeply affected by the Web changes as more formal elements predominate in these phases, particularly in the latter two. Nevertheless, some interesting experiments are showing the path to a more innovative approach both in research certification and evaluation. With regard to certification, experiments of open peer-review are being carried out by some e-journals (e.g., PLoS ONE, Atmospheric Chemistry and Physics) or have been performed in the past years (Nature's 2006 trial of open commentary) and are challenging the consolidated single-blind and double-blind peer-review system. With regard to the research evaluation, new approaches are based on usage metrics (Journal Usage Factor). These innovative measures integrate with the more traditional citation based ones (i.e., the Impact Factor, H-index). Yet "these initiatives still lack the necessary institutional awareness" (**Ponte & Simon, 2011**, p. 150) and do not substitute the citation metrics.

The full potential and the innovative communication models offered by the technology, particularly the combination of the Open Access/Open Data paradigm with the most advanced tools of the Web 2.0, are also reconfiguring the traditional relationship between science and society and provide new impetus to the concept of "citizen science". Although this collaborative model has a long history and is certainly more suitable for some domains and types of research (**Lyon, 2009**), some consolidated and successful examples of citizen science can be given in different disciplines, e.g., in astronomy (GalaxyZoo1 andStardust@ home2), meteorology (ClimateWatch3) and biology (Rosetta@home4).

Notwithstanding this increasing consolidation of Web 2.0 technologies in the scholarly communication, it is still difficult to understand whether or not this remains a niche phenomenon with generational and disciplinary biases. It is also important to know whether social media bring some benefits to the research and teaching workflow, and if so, what kind of benefits.

1.2 Web 2.0 technologies in research and teaching workflow: a literature review

To date, many articles have generically dealt with the topic of Web 2.0 in the scholarly communication and research workflow (**Kalb et al., 2009**; **Mahapatra, 2010**; **McMahon et al., 2012**). More specifically, the topic of the adoption of the Web 2.0 tools by academic research communities has already been explored in a few surveys conducted in different countries. One of the first studies on the way in which researchers are making use of Web 2.0 tools in the course of their researches was carried out by the Research Information Network in the UK in 2009 (**RIN, 2010**). The study developed a composite methodology combining an online survey with an in-depth, semi-structured interview with a sample of 56 survey respondents, including a total of fifteen semi-structured interviews with service developers and twenty interviews with Web users. 1,282 valid responses were obtained from the online survey. Findings show that 45% of the respondents are occasional users of Web 2.0 tools. Researchers tend to use mostly well-known generic tools such as Google Scholar (73%) and Wikipedia (69%) while a significant minority also use social networking services such as YouTube (29%), Facebook (24%) and Twitter (10%). Overall, however, the RIN study highlights a low level of uptake among the UK research community of Web 2.0-based services.

In 2010, another relevant study on the use and impact of Web 2.0 tools on research workflows was carried out by CIBER at UCL and funded by the Emerald Publishing Group. The CIBER study was carried out internationally online and received 1,923 complete answers from academics. Findings are aligned with the RIN study as the results show that the most established Web 2.0 tools are also the most popular ones: tools of collaborative authoring (e.g., Google Docs) are by far the most popular (62.7%), followed by social conferencing tools (e.g., Skype, 48.3%) and by scheduling and meeting tools (e.g., Google Calendar and Doodle). In order to share images and videos, 69% of respondents used YouTube, 14% used SlideShare and 12% Flickr. The preferred bookmarking service was Delicious. The most used social bookmarking platform was Facebook. The CIBER study also highlights subject differences in the use of the Web 2.0 technologies: natural and computer scientists were the most frequent users of social media, while social scientists and humanities researchers, albeit attracted by the new communicative tools, stay behind. With regard to age, there is a statistically clear distinction between researchers under- and over-35 years old. Yet "it is very difficult to detect any general overall pattern" (Rowlands et al., 2011, p. 188).

A third survey on the adoption of Web 2.0 technologies in the research workflow was carried out by **Ponte and Simon (2011)** among researchers of different disciplines in Europe from May to August 2010. The authors obtained 345 full responses. Findings show that researchers massively used search engines (Google Scholar was used by 99.7% of respondents). Among Web 2.0 tools, wikis (42%), blogs (38.6%) and social networks (34.8%) are fairly popular. Social bookmarking (25.8%) and micro-blogging (17.7%) are less used.

A similar Web survey was conducted in Finland in November 2009 by **Gu and Widen-Wulff (2011)** among a targeted sample of Abo Akademi University academic staff. Findings show that researchers are well-acquainted with Web 2.0 tools. However, "respondents use more multimedia sharing and social networks in everyday life than in research or teaching work" (**Gu & Widen-Wulff, 2011**, p. 768). Researchers tend to use different social tools according to their scope in research.

Scholars from educational, social and behavioural sciences such as **Alexander (2006)**, **Anderson (2007)**, **Churchill (2009)** and later **Churchill, Wong, Law, Salter, and Tai (2009)** and **Grosseck (2009)** started analyzing the use of Web 2.0 technologies in educational activities. All the above-mentioned researches stress the potential of the Web 2.0 to promote a deeper learning and foster the responsible adoption of social technologies in teaching activities. In particular, Churchill et al. describe the experience of the Repository of Interactive Social Assets for Learning (RISAL), the social platform implemented in 2009 at the University of Hong Kong which integrates social bookmarking, repository and networking and supports students and teachers in managing, sharing and reusing learning and teaching resources.

In 2008, in order to gauge the level of acceptance of Web 2.0 applications in learning to "inform educators in higher education for engaging and effective learning experiences", **Yoo and Huang (2011)** conducted a survey among American and Korean college students. Their results show that Korean students massively use blogs and online social communities but suffer from a high level of anxiety in using online conferencing tools (e.g., Skype) and social virtual environments (e.g., Second Life), while American students do not use as many Web 2.0 tools as Korean students do but find themselves at ease in online social communities. Korean students also feel less at ease in using Web 2.0 tools for learning; they admit that Web 2.0 technologies are useful in learning, but in comparison with American students they do not use them. These results are aligned with **Cavalli (2012)** who gives evidence to the fact that for digital natives a high level of use of Web 2.0 technologies for personal needs is not necessarily predictive of a high use of the same technologies for learning activities.

Another stream in the literature focuses more directly on the application of a single Web 2.0 technology in the education environment, in particular wikis (**Chao, 2007**; **Parker & Chao, 2007**), blogs and microblogs (**Churchill, 2009**;**Ebner & Maurer, 2009**).

To date, no survey has explored the use of social media in both research and teaching activity. This was the first goal of our study. Moreover, we were intrigued by the rapid change in the modes of communication and in the use of Web 2.0 technologies. It was, therefore, interesting for us to explore if, two years after the last survey, there had been relevant changes in the adoption of Web 2.0 technologies in research and learning activities. Finally, as impact of learning and using the technology is influenced by culture-related aspects5 (**Collis, 1999**), we decided it was interesting to us to conduct a survey on the use of social media in a well-developed and technology-oriented country like The Netherlands.

2. METHODOLOGY AND FINDINGS

Many definitions of Web 2.0 can be found in the literature (**O'Reilly, 2008;Siemens, 2008**). For the purpose of our survey, we decided to adopt Anderson's Web 2.0 definition: "Web 2.0 encompasses a variety of different meanings that include an increased emphasis on user-generated content, data and content sharing and collaborative effort, together with the use of various kinds of social software, new ways of interacting with web-based applications, and the use of the web as a platform for generating, re-purposing and consuming content" (**Anderson, 2007**).

The survey was conducted in April 2012 through an online questionnaire, which was divided in three sections:

1. personal information
2. Web 2.0 tools and their adoption in research activities
3. Web 2.0 tools and their adoption in teaching activities.

The questionnaire was sent by internal mail to university lecturers of the Media and the Game departments of the Breda University of Applied Sciences in The Netherlands. Out of approximately 60 addressees, 29 accessed the online survey and 12 completed it. Six addressees started but did not finish it. Most of them stopped after completing the first demographics questions, right before the survey started to ask more specific details about their social media usage. As a result, only 12 questionnaires were analysed.

The final respondents were either at the level of junior and senior lecturer/researcher and, in terms of research discipline, they were heterogeneously and not evenly distributed among the fields of arts and humanities (2 respondents), social sciences (4 respondents), computer science (5 respondents) and business, marketing and management (1 respondent). Age varied considerably, with the majority being between 41 and 45 years.

The survey questions can be found in Appendix A. Below, replies are grouped according to the two main topics that were addressed by the enquiry and will be discussed in detail in the following two subsections.

2.1 The adoption of Web 2.0 tools for research activities

For research activities, most respondents (n=8) use Web 2.0 tools regularly for their professional activities, 3 respondents use them, but only rarely, and only 1 respondent stated to use them solely for personal, private purposes. The most used social networking platform is LinkedIn as indicated by almost all respondents (n = 10), while Facebook is used only by 6 respondents. For this question multiple answers were possible: this explains why the total amount of answers exceeds the number of accepted respondents (that is those who completed the questionnaire), although 2 of the registered 12 did not indicate any social platform. Because the survey was filled out anonymously online, we were not able to investigate further how and for which research activities such social platforms are actually used. Besides these platforms, Wikis and blogs seem to be rather popular: 9 respondents declare to be using Wikipedia and 5 use institutional Wikis, whereas scientific blogs are used by 5 of the respondents, with 3 of them using their personal blog and only 1 the personal blog of one of their colleagues. These were also questions for which multiple answers were possible. Microblogging platforms are used by a very restricted number of respondents who are all using Twitter, with one respondent indicating specifically to be using the Twitter account of the university.

Social bookmarking or reference management tools are not that well known, according to the number of answers received to this question.

Only 3 respondents replied, each of whom indicating a different answer: one using Mendeley, one Zotero and one CiteULike.

Collaborative project platforms are not used at all, maybe even not known, given that only one person filled out this question. We fear that the term was not recognised. This single reply, in fact, marked "Others" specifying that the platforms used are Dropbox and N@tschool, which is the e-learning platform developed by the University of Breda. Other kinds of collaborative platforms include all Google-related products, i.e., Google Calendar, Google Docs, GTalk, along with Skype and YouTube as indicated in Figure 1.

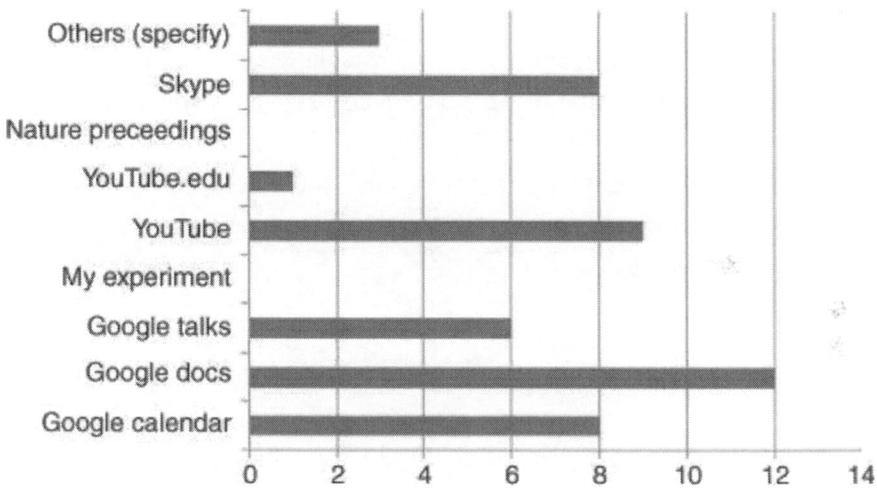

Figure 1. Collaborative platforms and tools used.

When we look at the reasons why adopting Web 2.0 tools is considered positive from a research point of view, respondents indicate that such tools:

- are free to use (mentions: 100%);

- are easy to use as there is only a minimum set of skills required to be able to use them (11 mentions);
- help them keep in touch with their colleagues (9 mentions);
- help them work in a collaborative way (8 mentions), with only 1 respondent not agreeing with this statement;
- help them share and disseminate fast ideas and research results (6 mentions);
- help them to keep updated in their own research field (6 mentions); however, four respondents claim that Web 2.0 tools do not help them at all in remaining up-to-date in their research;
- help them save time and costs (i.e., travelling is less necessary....). 6 respondents find that Web 2.0 can be used as a replacement for personal travels because of the possibility to have conference calls, of sharing resources remotely as indicated earlier, or of communicating synchronously. However, 3 respondents do not see these possibilities.

At the other side of the spectrum, the reasons for not adopting Web 2.0 tools in their research activity are:

1) the fact that the use of Web 2.0 tools conceals a sum of technologies and concepts which are still insufficiently defined; however, the majority of the respondents (8 mentions = 67%) disagree with this statement;
2) being very busy and finding that using these tools takes too much time; this answer was indicated by only 3 respondents; most respondents (9 mentions) disagree with this statement;
3) the low quality of the content shared; only 2 respondents agree with this statement;

4) privacy concern: only indicated by 1 respondent;

5) the fact that Web 2.0 tools promote amateurishness by opening contents to non-academic users was not claimed by anybody: no single respondent believes that the democratic character of the Web has opened up the doors to non-professionalism;

6) not trusting Web 2.0 tools and platforms: all respondents do not agree with this statement;

7) the fact that in their own research field collaboration is not a modus operandi (so researchers mainly work by themselves); all respondents disagreed.

2.2 The adoption of Web 2.0 tools for teaching activities

The situation is somewhat similar when we look at the teaching activity. Nine of our respondents declare to use Web 2.0 tools regularly in their teaching activities, and only 3 use them seldom. No one among our respondents states not to use them at all, neither for professional nor for personal goals, or to use them only for personal goals.

Among the tools that are mentioned, we find many Google products, Dropbox, blogs, YouTube, CiteUlike and Moodle. Surprisingly, also Skype and Twitter are used as a teaching tool.

When asked for the purpose of use, the respondents indicated:

1) for communication and information sharing;

2) for giving, submitting and grading of assignments;

3) for collaboration and discussion with and among students;

4) for polls, examples, networking, presentations and viewing of streaming videos such as tedtalks6, film clips;

5) for referencing to Web sites or any other study material that are used as showcases for students' projects.

Table 1 shows how the respondents assessed the main advantages of Web 2.0 tools for teaching. As can be seen, the possibility of posting teaching resources (videos, slides, etc.) received most positive mentions (definitely agree – 2, agree – 9, neutral – 1). The immediate feedback from students is regarded nearly as positive. Nine respondents agree or definitely agree with this statement.

With regard to the main reasons for not adopting Web 2.0 tools in their teaching activity, respondents indicate:

1) lack of time (5 mentions),

2) lack of expertise (3 mentions),

3) nobody finds that there are privacy concerns involved in the use of Web 2.0 tools.

3. DISCUSSION OF THE RESULTS

According to the results of our survey, full awareness of the benefits of Web 2.0 tools in research activity is still to be acquired while the use of Web 2.0 technologies in teaching activities appears to be consolidating.

3.1 Social networks

LinkedIn is increasing its popularity and has become the most used social networking platform for professional activities, even exceeding the popularity of Facebook. This result highlights a change in the researchers' use of social media as they are beginning to use Web 2.0 tools that are more professional-tailored at least in the category of social networks. This is in contrast with what was reported by CIBER: "Researchers seem to be largely appropriating generic tools rather than using specialist or custom-built solutions" (**CIBER & Emerald Group Publishing, 2010**). Still Facebook is by far the most popular social network among students. This should be kept in mind by educators when choosing a platform to reach out to their students. In particular, they should consider the impact of the Network Effect. This is a generic term "used to describe the increase in value to the existing users of a service in which there is some form of interaction with others, as more and more people start to use it. [....] This network effect is driving the continual improvement of Web 2.0 services and applications as part of the architecture of participation" (**Anderson, 2007**, p. 20)6. The network effect has both social and economic implications. It explains why faculty find it easier and more effective to interact with students on Facebook rather than to aggregate them in the institutional Virtual Learning Environment (**Anderson, 2007**, p. 21).

3.2 Wikis

Wikipedia is by now a well-established and reputed reference resource. **Park (2011)** also gives evidence to the fact that citations to Wikipedia are increasing quickly in scholarly publications. Overall, wiki-style technology is becoming very popular in the research workflow and more and more integrated in the institutional research environment as 50% of our respondents declared to be using institutional wikis for research purposes.

Wikis in education can be used to accomplish different objectives. According to**Parker and Chao (2007)**, educational wikis support two learning approaches: the constructivist paradigm and the cooperative/collaborative learning paradigm. The latter is particularly effective if the wiki refers to a community of practice.

Table 1. Advantages of Web 2.0 tools for teaching.

	Totally disagree	Disagree	Neutral	Agree	Definitely agree
Immediate feedback from students	0	1	2	8	1
Posting of teaching resources (video, slides, etc.)	0	0	1	9	2
Creation and sharing of bibliography with students	1	1	4	5	1
Creating a more accessible, portable, durable and interactive educational portfolio	0	1	3	7	1
Creating a very good classroom environment	0	1	4	7	0
Better identifying students' interests and the use of teaching resources	0	2	6	4	0
Helping students to develop capabilities in communication and collaborative works	0	0	3	9	0

Although wikis can have manifold applications in education, they are particularly well-suited to support collaborative writing: "a wiki as a writing tool maximizes the advantages of reflection, reviewing and publication, and of observing cumulative written results as they unfold" (**Parker & Chao, 2007**, p. 61). Wikis are also frequently used for project based learning and documentation and for building collaborative bibliographies.

Unfortunately, as the scope of our survey was general, we were not able to investigate deeper how and for which learning activities wikis are actually used at the University of Breda.

3.3 Blogs

Blogs are used for different purposes in the research lifecycle: to disseminate research results, to identify research opportunities and collaboration, to review the literature, to collect research data. (**CIBER & Emerald Group Publishing, 2010**.)

Blogs were originally designed to support personal diaries. They differed hugely from wikis as wikis were conceived as multiple collaborative tools. Notwithstanding this, only 3 out of 12 respondents publish a personal blog while 42% of respondents declared to post comments on scientific blogs (e.g., ScienceBlog.com, Nature.com Blogs, Research Blogging, RealClimate). It can be a very time-consuming task indeed for researchers to publish regularly in a blog, but we presume that this result can be better explained by the fact that the blogosphere is increasingly becoming a relevant tool in the dissemination of new ideas, and blogs increasingly form a powerful social community-building tool. As a matter of fact, "blogs provide considerable scope to widen the audience for scientific papers and to assist in the process of public understanding of science and research" (**Anderson, 2007**, p. 35).

Blogs are also more and more integrated in scholarly publishing (Nature.com Blogs, the PLoSONE blog's EveryONE, BioMed Central blogs) and, therefore, they are more naturally conceived by researchers as a collaborative tool for disseminating ideas.

Microblogging platforms (mainly Twitter) are still used by a very restricted number of respondents. This result is aligned with the CIBER UCL study findings and can be explained by the fact that microblogging tools are really a new kind of social media and have yet to reach their full take up among researchers. Recently, Twitter is increasingly used to disseminate comments during conferences and seminars.

Blogs in education are also widely used. As a matter of fact, blogging facilitates and contributes to students' learning. However, the extensive use of blogs by students requires facilitator's figures to stimulate them to take part in the coursework (**Churchill, 2009**). Hence, educators' personality and their capacity to involve students in the teaching and learning activities appear strategic.

3.4 Drivers

Ease of use and openness of Web 2.0 technologies7 are the two main reasons for adopting such tools in the research process. Keeping in touch with colleagues is the third important driver to the adoption of social media for research purposes. To date, these tools do not substitute but complement the more traditional ones (journals, repositories, search engines) for keeping up-to-date in the profession (50% stated that Web 2.0 tools supports their professional awareness).

The possibility to receive an immediate feedback from students, to share coursework and teaching resources and to help students develop capabilities in communication and collaborative works are the main drivers in the adoption of the Web 2.0 tools. Indeed, the importance of feedback to and from students for learning in progress has been emphasized as a relevant pedagogical intervention (**Salter, 2008**).

3.5 Obstacles

Respondents declared to trust social media; trust is one of the main characteristics of Web 2.0 and one of the key components of its success. To our surprise, respondents declared that they are not concerned about privacy. However, privacy and Intellectual Property Right are both relevant issues in Web 2.0. **Zimmer (2008)**, for example, by citing Elmer (2004), warns about the dangers of such an environment: "where the collection of personal information is a prerequisite of participation inevitably entrenches power in the hands of the technology designers". Privacy issues related to the Web 2.0, however, are much more complex, and boundaries between concern and not-concern are blurred. Kate **Raynes-Goldie (2010)**, for example, gives evidence of the fact that students are more worried about controlling access to personal information (social privacy) rather than about how social networks might use that information (institutional privacy).

Lack of time is also not considered to be an obstacle when the use of social technologies refers to the research activity, but it may represent a drawback for their adoption in teaching and learning activities.

4. CONCLUSIONS

The concepts, projects, and practices of Web 2.0 as a whole, insofar as we have surveyed them, are extremely fluid.

In research activities, frequent use of social media is rare; only the use of LinkedIn is significant. The selection and uptake of Web 2.0 tools is mainly dominated by the Network Effect, and technologies are subject to alternating dooms.

In teaching and learning, the use of Web 2.0 technologies is consolidating, both at the personal and institutional level, and it presents interesting perspectives. Nevertheless, educators should choose Web 2.0 tools very carefully according to their teaching needs, course aims and personal attitudes. In teaching activities much more than in research, the adoption of social media should always be subject to a more general teaching planning to support the development of young people's skills in creativity and innovation and lifelong learning.

Although the majority of our respondents disagree with the statement that Web 2.0 technologies are not clearly defined, we derived the impression that terminology and differences among the Web 2.0 technologies and tools are neither clear nor consistent. Moodle or Dropbox are not *stricto sensu* Web 2.0 technologies, but they were cited by our respondents. This terminology confusion is also a logical consequence of the fact that Web 2.0 technologies are mainly in beta version and open source and are extremely innovative.

Finally, further and deeper studies are necessary to explore if and how social media support the critical thinking and conscious selection of information.

NOTES

[1]http://www.galaxyzoo.org/

[2]http://stardustathome.ssl.berkeley.edu/

[3]http://www.climatewatch.org.au/

[4]http://boinc.bakerlab.org/rosetta/

[5]Cultural aspects to be taken into account occur at different levels: societal, personal, organizational and disciplinary.

[6]http:/www.ted.com/talks

[7]The drawback of the Network Effect is the risk to become 'locked in' to a product.

[8]Openness is indeed one of the main characteristics of the Web 2.0 and is a prerequisite of the collaborative paradigm.

REFERENCES

1. Alexander, B. (2006). Web 2.0: A new wave of innovation for teaching and learning? EDUCAUSE Review. 41, 32–44. Retrieved

September 17, 2012, from: http://www.educause.edu/ero/ article/ web-20-new-wave-innovation-teaching-and-learning.

2. Anderson, P. (2007). What is Web 2.0? Ideas, technologies and implications for education. Technical report, JISC. Retrieved September 17, 2012, from: http://www.jisc.ac.uk/media/ documents/techwatch/tsw0701b.pdf.

3. Bukvova, H. (2010). Studying research collaboration: a literature review. Sprouts: Working papers on information systems, 10 (3). Retrieved September 17, 2012, from: http://sprouts.aisnet.org/10-3.

4. Cavalli, N. (2012). I nuovi alfabeti della biblioteca. Nativi digitali: nuovi alfabeti per i lettori, Conference, Milan, Italy: March 15–16.

5. Chao, J. T. (2007). Student project collaboration using Wikis. Proceedings of the 20th conference on Software Engineering Education & Training (CSEET 2007), pp. 255–261. Dublin, Ireland: July 3–5. doi: 10.1109/CSEET.2007.49.

6. Churchill, D. (2009). Educational applications of Web 2.0: using blogs to support teaching and learning. British journal of educational technology, 40 (1), 179–183. Retrieved September 17, 2012, from: http://ayersresources.wikispaces.com/file/ view/ Educational+applications+of+Web+2.0+-+Using+blogs+to+support+ teaching+and+learning.pdf.

7. Churchill, D., Wong, W., Law, N., Salter, D., & Tai, B. (2009). Social bookmarking-repository-networking: possibilities for support of teaching and learning in higher education. Serials review, 35 (3), 142–148. doi: 10.1016/j.serrev.2009.04.006.

8. CIBER & Emerald Group Publishing (2010). Social media and research workflow. Retrieved February 18, 2013, from: http://ciber-research.eu/download/20101111-social-media-report.pdf.

9. Collis, B. (1999). Designing for differences: cultural issues in the design of WWW-based course-support sites. British journal of educational technology, 30 (3), 201–215.

10. Ebner, M., & Maurer, H. (2009). Can weblogs and microblogs change traditional scientific writing? Future Internet, 1, 47–58. Retrieved September 17, 2012

11. Finholt, T. A. (2003), Collaboratories as a new form of scientific organization, Economics of innovation and new technology 12 (1), 5–25.

12. Grosseck, G. (2009). To use or not to use web 2.0 in higher education? Procedia social and behavioural sciences, 1, 478–482. Retrieved September 17, 2012, from: http://webpages.csus.edu/~sac43949/pdfs/to%20use%20or%20not%20to%20use.pdf.

13. Gu, F., & Widén-Wulff, G. (2011). Scholarly communication and possible changes in the context of social media. A Finnish case study. The electronic library, 29 (6), 762–776.

14. Kalb, H., Bukvova, H., & Schoop, E. (2009). The digital researcher: exploring the use of social software in the research process. Sprouts: Working papers on information systems, 9 (34). Retrieved September 17, 2012, from: http://sprouts.aisnet.org/9–34.

15. Lyon, L. (2009). Open science at web-scale: Optimising participation and predictive potential consultative report. DCC Report commissioned by JISC. Retrieved September 17, 2012, from: http://www.jisc.ac.uk/publications/reports/2009/opensciencerpt.aspx.

16. Mahapatra, A. (2010). Researchers in a brave new Web 2.0 world. ACS Chemical biology, 5 (9), 799–800.

17. McMahon, T. M., Powell, J. E., Hopkins, M., Alcazar, D. A., Miller, L. E., Collins, L., et al. (2012). Social awareness tool for science research. D-Lib Magazine, 18 (3/4). Retrieved September 17, 2003.

18. O'Reilly, T. (2008). Why Dell.com (was) more Enterprise 2.0 than Dell IdeaStorm. 29 September. Retrieved September 17, 2012, from: http://radar.oreilly.com/2008/09/why-dell-dot-com-is-more-enterprise.html.

19. Park, T. K. (2011). The visibility of Wikipedia in scholarly publications. First Monday, 16 (8). Retrieved September 17, 2012.

20. Parker, K. R., & Chao, J. T. (2007). Wiki as a teaching tool. Interdisciplinary journal of knowledge and learning objects, 3, 57–72. Retrieved September 17, 2012, from: http://ijklo. org/ Volume3/ IJKLOv3p057-072Parker284.pdf.

21. Ponte, D., & Simon, J. (2011). Scholarly communication 2.0: exploring researchers' opinions on Web 2.0 for scientific knowledge creation, evaluation and dissemination. Serials review, 37 (3), 149–156.

22. Raynes-Goldie, K. (2010). Aliases, creeping and world cleaning: understanding privacy in the age of Facebook. First Monday, 15 (1). Retrieved September 17, 2012, from: http://firstmonday.org/htbin/cgiwrap/bin/ojs/index.php/fm/article/view/2775/2432.

23. RIN (2010, July). If you build it will they come? How researchers perceive and use web 2.0. Retrieved September 17, 2012, from: http://rinarchive.jisc-collections.ac.uk/our-work/communicating-and-disseminating-research/use-and-relevance-web-20-researchers.

24. Salter, D. J. (2008). The challenge of feedback: too little too late. In J. Luca & E. Weippl (Eds.), Proceedings of World conference on educational multimedia, hypermedia and telecommunications 2008 (pp. 3925–3926). Chesapeake, VA: AACE.

25. Siemens, G. (2008). A world without courses. 6 March, 2008. Elearnspace (blog). Retrieved September 17, 2012, from: http://www.elearnspace.org/blog/2008/03/06/a-world-without-courses/.

26. Yoo, S. J., & Huang, W. D. (2011). Comparison of Web 2.0 technology acceptance level based on cultural differences. Educational technology & society, 14 (4), 241–252. Retrieved September 17, 2012, from: http://www.ifets.info/ journals/ 14_4/ 20.pdf.

27. Zimmer, M. (2008), The externalities of search 2.0: The emerging privacy threats when the drive for the perfect search engine meets Web 2.0. First Monday, 13 (3). Retrieved September 17, 2012, from: http://www.firstmonday.org/htbin/cgiwrap/bin/ojs/index.php/fm/article/view/2136

Chapter 5

Working places, furniture and technology: Strategies of flexibility of university library buildings - the case of Bozen/Bolzano

Klaus Kempf

ABSTRACT

For centuries people have visited libraries to find information, and the practical needs of housing collections and accommodating readers have been the driving force behind library design. As technology advances of the past 20 years have made it possible for people to find information without entering a library building, some have even asked whether the bricks-and-mortar library is doomed to extinction. But today we know that this (bad) vision hasn't come true. Of course, it was imperative to rethink library services fundamentally and in this context also to rethink completely the functions of library buildings. In this ongoing process one term became more and more a magic word - "flexibility"!

When Gerhard Liebers, the doyen of building libraries in Germany, came back from a study tour to the United States in the early fifties, his experiences condensed in the term "flexibility". He was absolutely enchanted by the in those days modern and in Germany, at least in the field of library construction, still unknown *modular construction principle* (Liebers, 2002). Over the years the requirement of flexibility as an important aspect in the planning and briefing of new library buildings became more and more important. Today it is regarded as one of the most decisive, perhaps*the* basic principle of modern library buildings and in particular it got a new quality.

1. "FLEXIBILITY" IN HYBRID LIBRARY TIMES

At the moment libraries live through a period of transition and radical change. They have to deal with printed as well as digital materials, both of which with specific requirements in terms of the library building design. The library concept is changing to a so-called "hybrid library". However, this means more than the housing of a variety of different materials under one roof and the development of a new service concept. The advent of the networked digital world has not led to the "deserted library", abandoned by students and users, as had always been feared (Carlson, 2001). The opposite took place: We experience today the revival of the library as a place of learning and study (Engel & Antell, 2004).

The hybrid library asks for the following qualities:

- Learners have different preferences: different learning styles mean a need for study space varieties and "zoning", such as quiet private areas, and busy social areas with low-level seating and tables, and with refreshment facilities.

- Learning is more collaborative: a need for "collaboratories", such as group study tables and rooms with shared access to technology and presentation facilities.

- Technology is developing rapidly: a need for facilities for continual training and retraining of users in large classes, small groups and as individuals.

- Access to network resources is universal: a need for cable management for wired-up study places - both computer workstations and plug-in points for laptops (Corral & Brewerton, 1999).

Hybrid library features

It is assumed that learning will become more collaborative, learners have different preferences, technology will continue and network access to information will be imperative. Thus, in addition to providing ready access to large book and periodical collections, hybrid library features will include:

- group study areas, whether separate rooms or zoned, with ICT capabilities;

- varied, and zoned, study accommodation, and social space;

- training facilities and help desks to demonstrate systems, services and techniques;

- computer workstations and wired up reader places as well as in the case of the existing of a wireless LAN lap top dock-in facilities.

The design features that symbolise the libraries changing role from custodian to gateway, and which effectively bring together users and resources are:

- as few as permanent internal walls as possible,
- systematic zoning of areas for different activities,
- an extensive and accessible network and power grid to enable the connection of reader places,
- load bearing floors sufficient for book stacks throughout the building.

This leads us to the remarks of Faulkner Brown (Faulkner Brown, 1989) and Andrew McDonald (McDonald, 1997) that "flexibility or adaptability of space is a key criterion". Thus the modern, student-friendly university library is focused on the following qualities:

- as few as permanent internal walls as possible,
- systematic zoning of areas for different activities,
- an extensive and accessible network and power grid to enable the connection of reader places,
- load bearing floors sufficient for book stacks throughout the building.

2. THE LEARNING CENTRE IDEA

The new idea of a "hybrid library building" quickly turned into reality. The birthplace of the first representatives of this new generation of library buildings was Great Britain. There, in the first half of the 90s in Great Britain in consequence of the so-called Follett Report (1993), which was the result of a research study about the future of universities and university libraries in the UK, was born a new concept of a university library. The absolute necessity to save money led to the conviction that the bundling of learning resources and a more student centred university management could offer a way out of the deep financial crisis of the British universities. Students who pay for their study were considered as customers who should find all they need "under one roof". At the same time this could be a low budget solution. That was the right moment to shape a new idea of university libraries and for the establishment of so-called "Learning centres" or "Learning resource centres". One of the protagonists of this idea was G. Bulpitt and his project, the Adsett Center of the Hallam University of Sheffield which is one of the most outstanding examples (Bulpitt, 1999).

One important attribute of this new concept of a scholarly library and the library building is the respect for different "learning styles" and how they dynamically change from more independent to more group-based varieties and vice versa. Scott Bennett retains even that the "comeback" of the libraries could be eased by supporting collaborative learning by offering adequate learning facilities (Bennett, 2003).

The concept of the university library and the aspect of flexibility in the design and the building of the university library of Bozen

The motto of the Free University of Bozen is: small, but beautiful - focused on teaching and learning and internationally minded. The library is considered an integrated part of the university concept and recognized as "the heart of the university".

The library concept lives up to the principles of the university concept:

- (multimedia) information centre of the university and the region,
- innovative and internationally minded,
- a place to study and to learn

I'll try to verify whether and how the Bozen building lives up to the above-mentioned requirements of flexibility of a library building in times of the "hybrid library". But first of all let us take a look on the main characteristics of the new construction.

3. MAIN CHARACTERISTICS AND PARTICULARITIES OF THE NEW LIBRARY FACILITY

The creation of new spaces for the university library was part of the project to erect a completely new university site close to the historical centre of Bozen; therefore in the year 1998 a Europe-wide architecture competition was organized. The winner project provided the library facilities in the central building of the new campus.

This building is a "public building" in the literal meaning of the term - open to the university and open to the town. It forms a monumental portal between the town and the university. It is a rather introverted construction. In the architect brief it was required, taking into consideration the micro climate in Bozen, in summer one of the hottest towns in Italy, not to create a "greenhouse", i.e. no light-flooded constructions of glass and steel which we often find today in Central and Northern European countries. The philosophy of the design of the winning project respects this intention very well. The building forms a very closed and hermetic looking space, only interrupted by a line of windows on the east side of the facade. From the architects' point-of-view, it is supposed to resemble an ancient Roman house - from the outside very modest, even rejecting, inside offering a lot of open space around the atrium or more precisely around a central light well.

In the same building, on the ground floor are accommodated the student restaurant and the cafeteria; besides the entrance to the library - on the first floor - are located the student administration offices. Beneath the roof of the building the students can find the facilities of the computer and the language centre.

So the library is really situated at the crossing of all the central services and "under one roof" the student can do all the activities - with exception of attending lectures - which are particularly important for his daily life at the university. The library space itself (in the part accessible for the user) consists of a 4-storey facility which extends from the first floor up to the 4th floor of the building around a central light well; in the first basement, under the student restaurant is accommodated and not accessible for the user the (compact shelved) stacks.

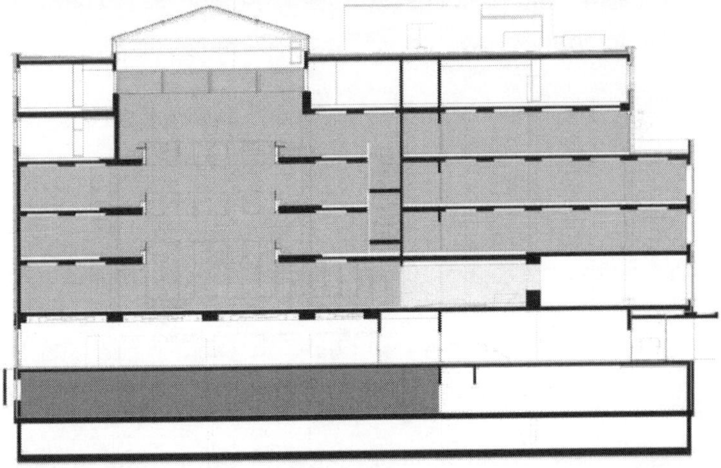

The library facilities are very spacious: The overall usable space amounts to 5.156 m2, of which 3.254 m2 are open access shelving (for at least 300.000 volumes) and reader places area (of which 865 m2 are lockable); the staff or office area (including the very spacious issuing and inquiry desk area) covers 317 m2 and the closed stacks area comprises

720 m2, enough to house 100.000 books as it was required in the architecture brief.

Let us take a short walk around the library: Starting on the ground floor where the main part of the space is occupied by the student restaurant and the cafeteria, which is also open to the public. Via a not very inviting staircase - this is definitively not a good solution to bring people into the library - we arrive on the first floor. Turning to the right hand side in a very large corridor and just in front of the entrance of the library we meet a very spacious and relaxing newspaper lounge; passing the exhibition and/or conference room we arrive at the (controlled) entrance to the library facilities with the central information/ inquiry/issuing desk on the right hand side and the cloakroom on the left hand side. The first floor or better the "ground floor" of the library facility houses also in a very compact way a cluster of staff offices.

In the midst of the building, the second and the third floor are structured in an identical way:

- On both levels, alternating with clusters of book stacks, we always find close to the windows 3 "islands" of user workplaces;

- On each of the two floors in the east corner of the building is located a cluster of carrels/small group study rooms and

- in the west corner of the building - on each of the two floors - a cluster of 4 very spacious seminar rooms

On the fourth level of the building around the library spaces are the facilities of the computer and the language centre. Also here in an alternating way we have the mixture of lockable spaces (a cluster of carrels) and the open access shelving and reader place area. Here is the "Silentium area" of the library with a lot of reader places primarily for the use of print material (and, if not noisy, laptops). Another attraction for users here on the top floor of the library facility is the opportunity to enjoy - thanks to a large and sunny balcony - open-air atmosphere.

4. VARIETY OF DIFFERENT WORKPLACES

First of all is to underline an extremely high number of user workplaces. The library offers altogether 520 user workplaces; considering the number of students at the Bozen Campus - at the moment a little bit more than 1000 - the ratio between the number of potential users and the workplaces available is extremely favourable. The recommendation in Germany for library facilities in the Social sciences and Humanities provides a ratio of 1 (workplace) : 12 - 15 (potential users/registered students). In Bozen the situation is radically different. It sounds like a fairy tale: the ratio is 1 : 2!

Also the criterion, variety of different workplaces, lives up extraordinarily well to the general requirements given at the beginning of the statement. There is a real "family" of very different work- and reader places, supporting any learning and study style. There are:

- Reader places in the so-called newspaper lounge (in front of the library entrance) with the possibility of quick information and/or a relaxed break on easy seats.
- A cluster of computer-based workplaces in the "information area" close to the library entrance primarily dedicated to OPAC- and/or database researches and assisted by the staff in face-to-face-support.
- Not a user workplace in the strict meaning of the term, just to make browsing between the stacks and the shelved books more comfortable and to allow a quick look into the book, but an imperative in a fashionably furnished modern library building: the "pull-out reference shelf"
- The librarians were very sceptical, but even reader places which are very closely situated to the noise transmitting central light well, type bench, on the 2nd and 3rd floor are accepted and used as "intermediate sit-down-and reading-arrangements" by the patrons

- A chain of workplaces on one long table which is winding like a crown around the balustrade of the central light well on the 4th floor; every workplace is equipped with a desk lamp which can be switched on and off individually.

- Workplaces (with laptop docking possibility and equipped with individual desk lamps) zoned as "work place islands" between the bookstacks in the open access area on the 2nd and 3rd floor.

- A variety of *lockable workplaces*: the 2nd and 3rd level of the building are structured, as just mentioned, in an identical way. Here we have a mixture of open access areas and lockable spaces. More than 800 m^2 of the overall existing usable space of the building used for library purposes is lockable. Lockable space means a more individual use of space, in particular it allows, even promotes "team work" amongst the students. To create as much group based working facilities as possible was also one of the key requirements in the architect brief. The main part of these rooms has windows and they have quite good daylight.

The lockable spaces are arranged as clusters. There are three categories of lockable spaces:

- *One and more person carrels (so-called "studies")*: the best privacy and the most intimate atmosphere are guaranteed by the carrels or studies. The surface of a single carrel amounts to 9-11 m^2, which is rather spacious and without problems it could host also a little working group of 2, 3 or even 4 students. The surface of a two-person-cabin amounts to 11-13 m^2. The equipment standard comprises: table (docking possibility to the LAN); seat and bookshelf. As one can imagine, the carrels are heavily used.

- *Small group study rooms*: each surface amounts to approximately 31 m^2; equipped with tables (providing docking to the LAN), seats and a large book stack.

- *Seminar or so-called large group study rooms*: facilities with or without computer based workplaces; each of them comprises approximately 50 m^2. They are well equipped with filing cabinets, rolling seats, tables and presentation facilities. The facilities are not only used by students in the way of self-organized study group work, but also for user instruction (done by the library staff) and as lecture halls (by the faculty).

5. LIGHTING AND ACOUSTICS IN THE "HYBRID LIBRARY"

*Lighting*and *acoustics* are considered as some of the crucial points and often even as the major problems in modern library buildings, type student-centred library. The development of open-space libraries in general has increasingly led to *noise* problems, especially as it seems to coincide with the tendency for students to work in (often chatty) groups. Some users complain that group work noise undermines private study. However, we have to take into consideration that the phenomenon "acoustics-noise" is a very subjective feeling. Some readers even like an "active" or noisy social learning environment, whereas others have a preference for quiet study places with good acoustics and visual privacy.

The *lighting* problem is a double one. The problem is no longer to guarantee the user an optimal daylight reader place and to protect the deposited material as much as possible against paper destroying sunlight. Rather the problem of today is to develop an adequate lighting scheme, which satisfies the needs of the simple book reader who prefers a daylight workplace, and at the same time is computer- or rather screen-friendly and supports and promotes the use of computers in the library.

The library building of Bozen has a complicating lighting as well as acoustic situation. It is intended that the variety of space will allow students to find optimum working conditions on each visit: the type of study places which suits them, the right level of noise, and the appropriate lighting

5.1. Lighting situation in the Bozen building

The design of the Bozen university library tries to provide lighting, which meets both readers' needs and computer/screen requirements. The limited number of windows leads to the channelling and/or filtering of the daylight to certain parts of the building only. The latter is helpful for the zoning of the workplaces (classical reader places close to the windows, computer based work places put more distantly in the darker parts of the building). The central light well, covered by a glass roof, allows only the entry of filtered daylight. Only immediately beneath the glassed roof of the light well (on the 4th floor) there is plenty of sunlight and on the same floor the long balcony window front guarantees "sun bathed" reader places; on the 2nd and on the 3rd floor reader places situated immediately at the "well border" also get enough light. The rest of the library space is dived in a scattered light. Although this form of lighting is really screen friendly most of the users and in particular the staff prefer in addition also during the day hours a permanently switched on artificial (down-)light.

5.2. Anti-noise strategy used in the Bozen building

The concept of the library - in particular as student centred facility and the special characteristics of the design of the university library - force to think about an adequate anti-noise strategy. Primary noise sources in the library facility are the central staircase, at least the part connecting the 1st and the 2nd floor and the floor crossing central light well which works as a enormous "noise transmitter". While the problem of the staircase can be resolved only by radical intervention, it is erected badly and needs a complete reconstruction; the central light well is a basic element of the design concept and therefore cannot be eliminated.

In the struggle against noise we make use of "classical" remedies- to put a sound absorbing carpet on the floor and/or the use of noise absorbent surface materials under the ceiling and/or along the walls. But in open-plan buildings this normally isn't enough. In constructions like the Bozen library facility with its very open structure and the four storeys crossing central light noise cannot be extinguished only by a traditional anti-noise strategy.

Very helpful in the struggle with undesired noise is a sufficient depth of space on each floor. The further away from the noise sources, here the central light well, a workplace is situated, the quieter the workplace. In the case of Bozenthe depth of space on the 2nd and 3rd floor is sufficient to put work places at a minimum of 10m and more than 30m away from the central light well, which are relatively quiet and in the second case even silent; on the 4th floor the so-called "Silentium area" is more than 30m away from the light well balustrade.

A corrective can be also a consequent zoning - different user activities have been designated separate zones; zones delineates and physically separates various activities, notably the division between PC areas, quiet study areas and group study areas, and their respective acceptable levels of noise. Each zone accommodates different study space. The PC areas, for example, provide open-plan workstations where informal group work is permitted. By contrast, the quiet study areas, composed also of carrels, are essentially quiet reading areas for students using printed material and/or laptops. The principle should be that quiet study areas should be situated away from noisy high traffic and group study areas.

This policy was adopted in Bozen: to create a positive acoustic situation in the whole building led to the decision to locate acoustically problematic activities (and services) on the 1st floor, close to the entrance. Consequently, levels become quieter as one moves up the building; the top floor even has been set aside for quiet study and is called "Silentium area". On the various levels the block of book stacks act as a "sound barrier", absorbing noise. The highest level of noise protection offer the soundproof (individually used) carrels. But there is a problem: the dividing walls between carrels and the neighbouring so-called small group study rooms aren't soundproof, it's a simple (closed) wooden book stack construction.

6. FLEXIBILITY IN TECHNOLOGY - THE LOCAL AREA NETWORK (LAN)

A frequently expressed design objective is to wire-up as much space as affordable with network and power ports to maximise the building's convertibility but also, in part, to allow laptop computer use, although associated keyboard noise and need for sound proofing in quiet study areas is an important concern. Bazillon emphasises the need to accommodate the use of computers (including laptops) with network access built into study spaces. "Study spaces thus become self-contained work areas in which research and writing may be done with relative ease." (Bazillion & Braun, 1995).

The situation in Bozen is as follows:

- Only 48 networked and computer-based IT workstations are available (in lockable and open spaces), because the students find a considerable number of open accessible workstations in the faculties rooms and/or the computer centre facilities.

- Laptop plug-in or drop-in points are spread over the whole library facility.

- From all the 520 reader places (computer based or not) access to the (wired) LAN exists.

- The cavity floor permits a completely flexible cabling/wiring in the whole building.

- Since February 2004 also the "wireless" LAN is accessible in the library, but the wired LAN is still working, because it is much faster and has a higher security standard.

- The advantage of the wireless LAN - more economic (for the university) and more comfortable (for the students), because a plug-in to the LAN is also possible in areas where there is no cable (for instance, in the student restaurant and the cafeteria)

7. FLEXIBILITY IN FURNISHING

Furnishing forms an integrated part of any library building concept. There is often a controversial debate between architects and librarians about the various aspects of furnishing. But a compromise should be found. The general motto: "The more functional and the more flexible the better for librarians and the users" can also be applied here. The aspect of flexibility in furnishing means in particular the usage of standardized products, which allow the library an ease change or extension. Especially designed furniture, in particular book stacks and work place furniture should be the absolute exception and in general avoided. In the case of Bozen this "golden principle" was taken in consideration for the choice of the book stacks as well as for the work place furniture:

Book stacks: in principle all spaces in the building can be used for all library purposes. The load capacity of the (cavity) floor construction is sufficient to put the book stacks (even compact stacks) in any part of the open access area. The installed book stacks (in the open access area) are standard products and can be handled easily. Only in and outside of the carrels and/or small group study rooms are (wooden) book stack constructions made by a cabinetmaker and not so easy to move.

Work place furniture (in the user and staff areas): seats and desks are standardized products. They can be rearranged and/or changed without problems.

8. CONCLUSION

In life dreams rarely come true, and even less so in library building. In the construction and design of the university library of Bozen it was attempted to transform into reality as much as possible of the conceptual ideas i.e. *the service-centred library,* which had inspired the foundation and organisation of the new library from the very beginning. Given a first draft, which only partly reflected the librarians' wishes, eventually a compromise acceptable to both sides was reached after an intensive dialogue with the architects. The result is quite remarkable, especially as regards the library as a place of learning. The library rooms are very well accepted by its users, in particular the students. Or to put it in a nutshell: not a perfect world but a highly functional and convenient learning environment.

REFERENCES

1. Bazillion, R.J. and C. Braun: *Academic libraries as high tech gateways: a guide to design and space decisions*. Chicago [u.a.]: American Library Association, 1995, p. 44
2. Bennet, S.: *Libraries designed for learning*. Washington, 2003. Pp. 16-19.
3. Bulpitt, G.: "The Adsetts Learning Center, Sheffield Hallam University". In: Intelligent Library Buildings. Proceedings of the 10th Seminar of the IFLA Section on Library Buildings and Equipment. The

Hague, Netherlands, 24-29 August 1997. München: Saur, 1999, Pp. 245-259.

4. Carlson, Scott: "The Deserted Library: As Students Work Online, Reading Rooms Empty Out, Leading Some Campuses to Add Starbucks". *Chronicle of Higher Education*, (November 16, 2001). http:// chronicle.com/ free/v48/i12/ 12a03501.htm

5. Corral, S. and A. Brewerton: *The new professional's handbook: your guide to information services management*. London: Library Association, 1999, p. 214.

6. Engel, D. and K. Antell: "The life of the mind: A study of faculty spaces in academic libraries". *College & Research Libraries*, 65(2004)1, pp. 9-11.

7. Faulkner Brown, H.: "The open plan and flexibility". In: International Association of Technological University Libraries, Proceedings 11. London: IATUL, 1989, pp. 3-9.

8. *The Follett Report*. Bristol: Joint Funding Council's Libraries Review Group, 1993. http://www.ukoln.ac.uk/services/papers/ follett/ report/

9. Liebers, Gerhard: "Der Gedanke der 'flexibility' im neueren amerikanischen Bibliotheksbau". In: Liebers, Gerhard: Funktion und Gestalt der Bibliothek. Frankfurt an Mainz [u.a.]: Lang, 2002, 2, korrigierte Aufl., S. 9-26.

10. McDonald, A.: "Space planning and management". In: Baker, D.,ed.:*Resource management in academic libraries*. London: Library Association, 1997, pp. 189-206.

11. Powell, M.: "Designing library space to facilitate learning: A review of the UK higher education sector". *Libri*, 52(2002), p. 18.

Chapter 6

Lessons Learned: A Primo Usability Study

Kelsey Brett, Ashley Lierman, and Cherie Turner

ABSTRACT

The University of Houston Libraries implemented Primo as the primary search option on the library website in May 2014. In May 2015, the Libraries released a redesigned interface to improve user experience with the tool. The Libraries took a user-centered approach to redesigning the Primo interface, conducting a "think-aloud" usability test to gather user feedback and identify needed improvements. This article describes the method and findings from the usability study, the changes that were made to the Primo interface as a result, and implications for discovery-system vendor relations and library instruction.

1. INTRODUCTION

Index-based discovery systems have become commonplace in academic libraries over the past several years, and academic libraries have invested a great deal of time and money into implementing them. Frequently, discovery platforms serve as the primary access point to library resources, and in some libraries they have even replaced traditional online public access catalogs. Because of the prominence of these systems in academic libraries and the important function that they serve, libraries have a vested interest in presenting users with a positive and seamless experience while using a discovery system to find and access library information. Libraries commonly conduct user testing on their discovery systems, make local customizations when possible, and sometimes even change products to present the most user-friendly experience possible.

University of Houston Libraries has adopted new discovery technologies as they became available in an effort to provide simplified discovery and access to library resources. As a first step, the Libraries implemented Innovative Interface's Encore, a federated search tool, in 2007. When index-based discovery systems became available, the Libraries saw them as a way to provide an improved and intuitive search experience. In 2010, the Libraries implemented Serials Solutions' Summon. After three years and a thorough process of evaluating priorities and investigating alternatives, the Libraries made the decision to move to Ex Libris' Primo, which was done in May of 2014.

The Libraries' intention was to continually assess and customize Primo to improve functionality and user experience. The Libraries conducted research and performed user testing, and in May 2015 a redesigned Primo search results page was released. One of the activities that informed the Primo redesign was a "think-aloud" usability test that required users to complete a set of two tasks using Primo. This article will present the method and results of the testing as well as the customizations that were made to the discovery system as a result. It will also discuss some broader implications for library discovery and its effect on information literacy instruction.

2. LITERATURE REVIEW

There is a substantial body of literature discussing usability testing of discovery systems. In the interest of brevity, we will focus solely on studies and overviews involving Primo implementations, from which several patterns have emerged.

Multiple studies have indicated that users' responses to the system are generally positive; even in testing of very early versions by a development partner users responded positively overall.1 Interestingly, some studies found that in many cases users rated Primo positively in post-testing surveys even when their task completion rate in the testing had been low.2 Multiple studies also found evidence that, although users may struggle with Primo initially, the system is learnable over time. Comeaux found that the time it took users to use facets or locate resources decreased significantly with each task they performed,3 while other studies saw the use of facets per task increase for each user over the course of the testing.[4]

User reactions to facets and other post-limiting functions in Primo were divided. In one of the earliest studies, Sadeh found that users responded positively to facets,5 and some authors found users came to use them heavily while searching,6 while others found that facets were generally underused.7 Multiple studies found that users tended to repeat their searches with slightly different terms rather than use post-limiting options.8 Thomsett-Scott and Reese, in a survey of the literature on discovery tools, reported evidence of a trend that users reacted more positively to post-limiting in earlier discovery studies,9 while the broader literature shows more negative reactions in more recent studies. This could indicate that shifts in the software, user expectations, or both may have decreased users' interest in these options.

A few specific types of usability problems seem common across tests of Primo and other discovery systems. Across a large number of studies, it has been found that users—especially undergraduate students—struggle to understand library and academic terminology used in discovery. Some terminology changes were made after users had difficulty in the earliest usability tests of Primo,10 but users continued to struggle with terms like hold and recall in item records.11 Users also failed to understand the labels of limiters,12 and they also failed to recognize the internal names of repositories and collections.13 Literature reviews on discovery systems have found terminology to be a common stumbling block for searchers across a wide number of individual studies.14

Similarly, users often struggle to understand the scope of options available to them when searching and the holdings information in item records. Users failed in multiple tests to distinguish between the article level and the journal level,15 could not interpret bibliographic information sufficiently to determine that they had found the desired item,16 and chose incorrect options for scoping their searches.17 Many studies found that users were unable to distinguish between multiple editions of a held item when all item types or editions were listed in the record.18 In other cases, users had difficulty interpreting locations and holdings information for physical items.19

Among the needs and desires expressed by and for Primo users in the literature, two in particular stand out. First, many users expressed a desire for more advanced search options; some wanted more complexity in certain facets and the ability to search within results,20 while other users simply wanted an advanced search option to be available.21 Secondly, a large number of studies indicated that instruction on Primo or other discovery systems was needed for users to search effectively. In some cases this was the conclusion of the researchers conducting the study,22 while in other cases users themselves either suggested or requested instruction on the system.23

It is also worth noting that it has been questioned whether usability testing as a whole is a sufficient mechanism for evaluating discovery-system functionality. Prommann and Zhang found that usability testing has focused almost exclusively on the technical functioning of the software and not adequately revealed the ability of discovery systems like Primo to successfully complete users' desired tasks.24 They proposed hierarchical task analysis (HTA) as an alternative, to examine users' most frequent desires and the capacity of discovery systems to meet them. Prommann and Zhang acknowledged, however, that as HTA is completed by an expert on the system rather than by an actual user, some of the valuable information derived from usability testing (including terms and functions that users do not understand, however well-designed) is lost in the process; they concluded that a combination of the two systems of testing is ideal to retain the best of both.

3. BACKGROUND

At the University of Houston Libraries, the Resource Discovery Systems department (RDS) is responsible for the maintenance and development of Primo. However, it is important to RDS to gather feedback and foster buy-in from stakeholders in the Library before making changes to the system. To that end, RDS works with two committees to assess the system and make recommendations for its improvement. The Discovery Usability Group and the Discovery Advisory Group include members from public services, technical services, and systems; each member brings a unique perspective on discovery. The Discovery Usability Group is charged with assessing the discovery system through a variety of methods including usability testing, focus groups, and user interviews. The Discovery Advisory Group reviews results of user testing and makes recommendations for improvement. All changes to the discovery system are reviewed by the Groups before they are released for public use.

In fall 2014, several months after the Primo implementation, the Discovery Usability Group conducted a focus group with student workers from the library's information desk (a dual reference and circulation desk) to solicit feedback about the functionality of Primo and suggestions for its improvement. In the meantime, the Discovery Advisory Group was testing Primo and evaluating Primo sites at peer and aspirational institutions. The groups used the information collected through the focus group and research on Primo to make recommendations for improvement. RDS has access to a Primo development sandbox, and many of the recommended changes were made in the sandbox environment and reviewed by the two groups prior to public release.

Changes to the search box can be seen in figure 1. Rarely used tabs were replaced with a dropdown menu to the right of the search box to allow users to limit to "everything," "books+," or "digital library." To increase visibility, links to "Advanced Search" and "Browse Search" were made larger and more spacing was added.

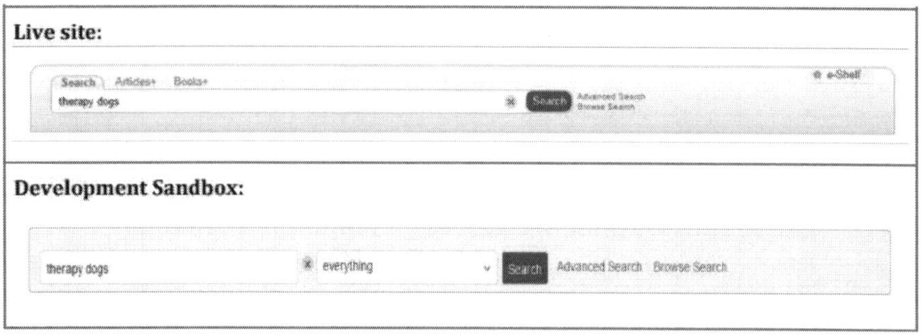

Figure 1. Search Box in Live Site (Above) and Development Sandbox (Below) at Time of Testing

Changes were also made to create a cleaner and less cluttered search results page (see figure 2). More white space was added, and the links (or tabs) to "View Online," "Request," "Details," etc., were redesigned and renamed for clarity. For example, the "View Online" link was renamed to "Preview Online" because it opens a box within the search results page that displays the item. The groups believed "Preview Online" more accurately represents what the link does.

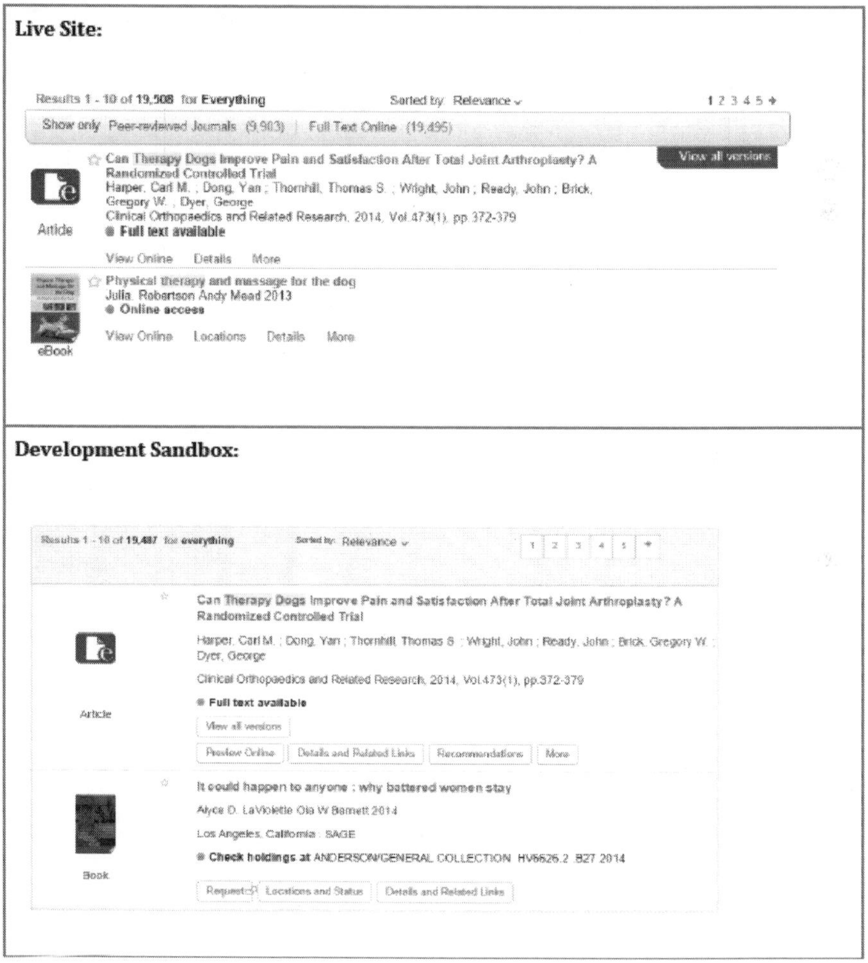

Figure 2. Search Results in Live Site (Above) and Development Sandbox (Below) at Time of Testing

The facets were also redesigned to look cleaner and larger to attract users' attention (see figure 3).

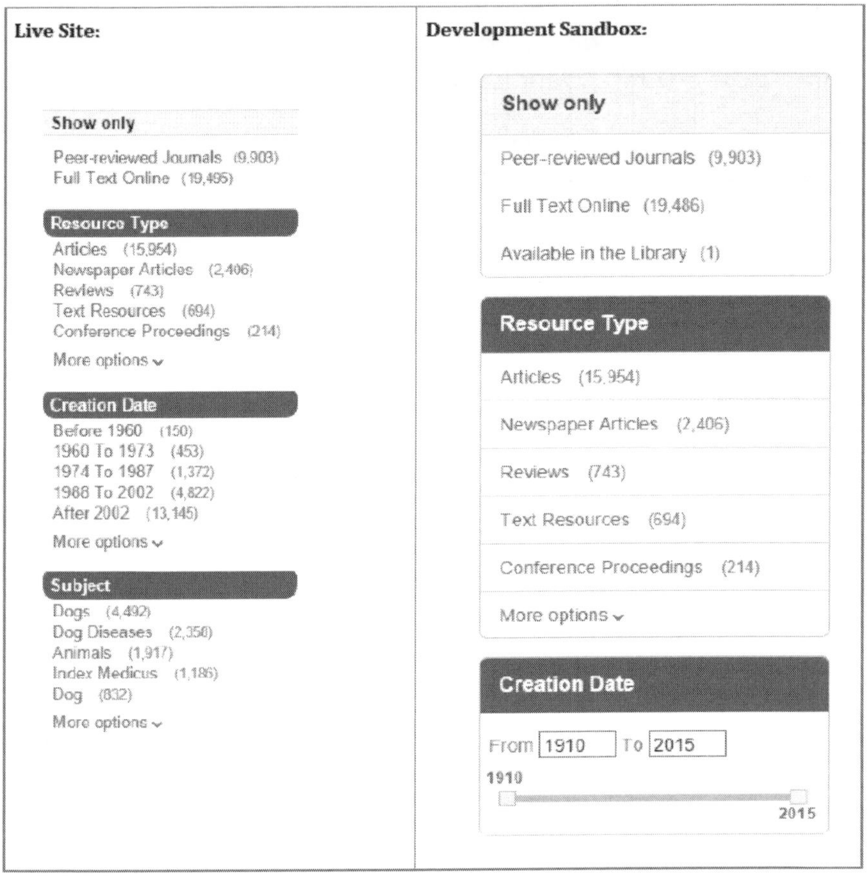

Figure 3. Facets in Live Site and Development Sandbox at Time of Testing

Both groups were happy with the changes to the Primo development sandbox but wanted to test the effect of the changes on user search behavior before updating the live site. The Discovery Usability Group conducted a usability test within the development sandbox. The goal of the test was to find out if users could effectively complete common research tasks using Primo. With that goal in mind, the group developed a usability test and conducted it during the spring semester of 2015.

4. METHODOLOGY

The Discovery Usability Group developed a usability test using a "think-aloud" methodology, where users were asked to verbalize their thought process as they completed research tasks through Primo. Four tasks were designed to mirror tasks that users are likely to complete for class assignments or for general research. To minimize the testing time, each participant completed two tasks, with the facilitators alternating between two sets of tasks from one participant to the next.

Test 1

Task 1: You are trying to find an article that was cited in a paper you read recently. You have the following citation:

Clapp, E., & Edwards, L. (2013). Expanding our vision for the arts in education. Harvard Educational Review, 83(1), 5–14.

Please find this article using OneSearch [the public-facing name given to the Libraries' Primo implementation].

Task 2: You are doing a research project on the effects of video games on early childhood development. Find a peer-reviewed article on this topic, using OneSearch.

Test 2

Task 1: Recently your friend recommended the book The Lighthouse by P. D. James. Use OneSearch to find out if you can check this book out from the library.

Task 2: You are writing a paper about the drug cartels' influence on Mexico's relationship with the United States. Find a newspaper article on this topic, using OneSearch.

Two facilitators set up a table with a laptop in the front entrance of the library. They alternated between the facilitator and note-taker roles. Another group member took on the role of "caller" and recruited library patrons to participate in the study. The caller set up a table visible to those passing by with library-branded T-shirts and umbrellas to incentivize participation. The caller explained what would be expected of the potential participant and went over the informedconsent document. After signing the form, the participant performed two tasks. After the test the participant received a library T-shirt or umbrella, and snacks.

The facilitators used Morae Usability Software to record the screen and audio of each test. Participants were asked for permission to record their sessions, but could opt out. During the three hour testing period, fifteen library patrons participated in the study, and fourteen sessions were recorded. Of the fifteen participants, thirteen were undergraduate students (four freshman, one sophomore, seven juniors, and two seniors), one was a graduate student, and one was a postbaccalaureate student. The majority of the participants were from the sciences, along with two students from the College of Business and two from the School of Communications. There were no participants from the humanities.

The facilitators took notes on a rubric (see table 1) that simplified the processes of coding and reviewing the recordings. After the usability testing, the facilitators reviewed the notes and recordings, coded them for common themes and breakdowns, and prepared a report of their findings and design recommendations. The facilitators sent the report, along with audio and screen recordings, to the Discovery Advisory Group, who reviewed them along with RDS. The Discovery Advisory Group made additional design recommendations, and RDS used the information and recommendations to implement additional customizations to the Primo development sandbox.

Table 1. Task Completion Rubric for Test 1

Preliminary Questions	
ASK: What is your affiliation with the University of Houston? Year? Major?	
ASK: How often do you use the library website? For what purpose(s)?	
Task 1	
Describe the steps the participant took to complete the task	S/U
ASK: How did you feel about this task? What was simple? What was difficult?	
ASK: Is there anything that would make completing this task easier?	
Task 2	
Describe the steps the participant took to complete the task	S/U
ASK: How did you feel about this task? What was simple? What was difficult?	
ASK: Is there anything that would make completing this task easier?	
Follow-up Question	
ASK: What can we do to improve the overall experience using OneSearch?	

5. RESULTS

Test 1, Task 1

You are trying to find an article that was cited in a paper you read recently. You have the following citation: Clapp, E., & Edwards, L. (2013). Expanding our vision for the arts in education. Harvard Educational Review, 83(1), 5–14.

Table 2. Results for Test 1, Task 1

Participant	Time on Task	Task Completion
1	1m 54s	Y
2	4m 13s	Y
3	1m 26s	Y
4	1m 17s	Y
5	1m 26s	Y (required assistance)
6	1m 43s	Y
7	1m 27s	Y
8	1m 5s	Y

All eight participants successfully completed this task, although sophistication and efficiency varied between participants. Some searched by the authors' last names, which was not specific enough to return the item in question. Four participants attempted to use advanced search or the drop-down menu to the right of the search box to pre-filter their results. Two participants viewed the options in the drop-down menu, which were "everything," "books+," and "digital library," and left it on the default "everything" search. When prompted, the participants explained that they were expecting the drop-down to contain title and/or author limiters. Similarly, participants expected an author limiter in the advanced search.

The citation format seemed to confuse participants, and they tended to search for the piece of information that was listed first—the authors—rather than the most unique piece of information—the title. If the first search did not return the correct item in the first few results, the participant would modify their search by searching for a different element of the citation or adding another element of the citation to the initial search until the item they were looking for appeared as one of the first few results. Participant 5 thought they had successfully completed the task, but the facilitator had to point out that the item they chose did not meet the citation exactly, and on the second try they found the correct item.

Participant 2 worked on the task for more than four minutes, significantly longer than the other seven participants. They immediately navigated to advanced search and filled out several fields in the advanced search form with the elements of the citation. If the search did not return their item, they added more elements until they finally found it. Simply searching the title in the citation would have returned the item as the first search result. Filling out the advanced search form with all of the information from the citation does not necessarily increase a user's chances of finding the item in a discovery system, though it might do so when searching in an online catalog or subject database.

The Discovery Advisory and Usability Groups made two recommendations to address some of the identified issues: include an author search option in the advanced search, and add an "articles+" option to the drop-down menu on the basic search. RDS implemented both recommendations. The Discovery Usability Group identified confusion around citations as a common breakdown during this task. The groups recommended providing instructional information about searching for known items to address this breakdown; however, RDS is still working on an effective method to provide this information in a simple and visible way.

Test 1, Task 2

You are doing a research project on the effects of video games on early childhood development. Find a peer-reviewed article on this topic, using OneSearch.

All eight participants successfully found an article on this topic, but were less successful in determining whether the article was peer-reviewed.

Only one participant used the "Peer-reviewed Journals" facet without being prompted. Three users noticed the "[Peer-reviewed Journal]" note in the record information for search results, and used it to determine if the article was peer-reviewed. One participant went to the full-text of an article, and said it "seemed" like it was peer-reviewed and considered the task complete. The resource type facets were more heavily used during this task than the "Peer-reviewed Journals" facet, despite its being promoted to the top of the list of facets. Two participants used the "Articles" facet, and two participants used the "Reviews" facet, thinking it limited to peer-reviewed articles. Participants 3 and 8 needed help from the facilitator to determine whether a source was peer-reviewed. There was an overall misunderstanding of what peer-reviewed means, which affected participants' confidence in completing the task.

Table 3. Results for Test 1, Task 2

Participant	Time on Task	Task Completion
1	3m 44s	Y
2	2m 21s	Y
3	5m 23s	Y (required assistance)
4	2m 5s	Y
5	3m 32s	Y
6	2m 45s	Y
7	3m 8s	Y
8	3m 1s	Y (required assistance)

The design recommendations based on this task included changing the "Peer-reviewed Journals" facet to "Peer-reviewed Articles" or simply,

"Peer-reviewed." RDS changed the facet to "Peerreviewed Articles" to help alleviate confusion. Additionally, the groups recommended emphasizing the "[Peer-reviewed Journal]" designations within the search results and providing a method for limiting to peer-reviewed materials before conducting a search. Customization limitations of the system have prevented RDS from implementing these design recommendations yet. A way to address the breakdowns caused by misunderstanding terminology also has yet to be identified. It was disheartening that participants did not use the "Peer-reviewed Journals" facet despite its being purposefully emphasized on the search results page.

Test 2, Task 1

Recently your friend recommended the book The Lighthouse by P. D. James. Use OneSearch to find out if you can check this book out from the library.

Table 4. Results for Test 2, Task 1

Participant	Time on Task	Task Completion
1	1m 7s	Y
2	56s	Y
3	No recording	Y
4	2m 21s	Y
5	1m 8s	Y
6	2m 14s	Y
7	1m 15s	Y

All seven participants were able to find this book using Primo, but had difficulty in determining what to do once they found it. For this task every participant searched by title and found the book as the first search result. Four users limited to "books+" before searching using the drop-down menu, while the other three remained in the default "everything" search. Only one participant used the locations tab within the search results to determine availability; the others clicked the title and went to the item's catalog record. All participants were able to determine that the book was available in the library, but there was an overall lack of understanding about how to use the information in the catalog to check out a book. Participant 1 said that they would write down the call number, take it to the information desk, and ask how to find it, which was the most sophisticated response of all seven participants. Participant 4 spent nearly two minutes clicking through links in the OPAC expecting to find a "Check Out" button and only stopped when the facilitator stepped in.

A recommended design change based on this task was to have call numbers in Primo and the online catalog link to a stacks guide or map. This is a feature that may be developed in the future, but technical limitations prevented RDS from implementing it in time for the release of the redesigned search interface. Like the previous tasks, some of the breakdowns occurred because of a lack of understanding of library services. Users easily figured out that there was a copy of the book in the library, but had little sense of what to do next. None of the participants successfully located the stacks guide or the request feature that would put the item on hold for them. Steps should be taken to direct users to these features more effectively.

Test 2, Task 2

You are writing a paper about the drug cartels' influence on Mexico's relationship with the United States. Find a newspaper article on this topic, using OneSearch.

This task was difficult for participants. Two users limited their search initially to "digital library" using the drop-down menu, thinking it would be a place to find newspaper articles; their searches returned zero results. Only two users used the "Newspaper Articles" facet without being prompted, and users did not seem to readily distinguish newspaper articles as a resource type. Participants did not notice the resource type icons without being prompted. Several participants needed to be reminded that the task was to find a newspaper article, and not any other type of article. With guidance, most participants were able to complete the task. Participant 4 remained on the task for almost eight minutes because of their dissatisfaction with the relevancy of the results to the prompt. Interestingly, they found the "Newspaper Articles" facet and reapplied it after each modified search, suggesting that they learned to use system features as they went.

Table 5. Results for Test 2, Task 2

Participant	Time on Task	Task Completion
1	4m 45s	Y (required assistance)
2	59s	Y
3	No recording	N
4	7m 47s	Y
5	2m 52s	Y
6	1m 33s	Y
7	1m 30s	Y

One of the recommendations based on this task was to remove "digital library" as an option in the drop-down menu on the basic search. It was evident that "digital library" did not have the same meaning to end users as it does to internal users. This recommendation was easily implemented. Another recommendation was to emphasize the resource type icons within the search results, but we have not determined a way to do so effectively. One suggestion from the Discovery Usability Group was to exclude newspaper articles from the search results as a default, but no consensus was reached on this issue.

6. LIMITATIONS

The Discovery Usability Group identified limitations to the usability test that should be noted. Testing was done in a high-traffic portion of the library's lobby, which is used as study space by a broad range of students. Participants were recruited from this study space, and we chose not to screen participants. The fifteen participants in the study did not constitute a representative sample. Almost all participants were undergraduate students, and no humanities majors participated. The outcomes might have been different if our participants had included more experienced researchers or students from a broader range of disciplines. By adding screening questions or choosing a more neutral location, we would have limited the number of participants who could complete our testing.

Another limitation was that the participants started the usability test within the Primo interface. Because Primo is integrated into the Libraries' website, users would typically begin searching the system from within the library homepage. The goals of the study required testing of our Primo development sandbox, which was not yet available to the public, and therefore could not be accessed in the same way. This gave participants some additional options from the initial search pages that are not usually available through the main search interface. While testing an active version of the interface would be preferable, one of our goals was to understand how our modifications affected user behavior, so testing the unmodified version was not an acceptable substitute. Additionally, the usability study presented tasks out of context and did not replicate a true user-searching experience. Despite the limitations, we learned valuable lessons from the participants in this study.

7. DISCUSSION

Users successfully completed the tasks in this usability study. Unfortunately, they did not take advantage of many of the features that can make such tasks easier—particularly facets. This was especially apparent when we asked users to find a peer-reviewed journal article (Test 1, Task 2). Primo has a facet that will limit a search to only peer-reviewed journal articles, and only one out of eight participants used this facet during this task. Participants appreciated the pre-search filtering options, and requested more of them (such as an author search), while post-search facets were underutilized.

Similarly, participants almost uniformly ignored the links, or tabs, within the search results, which would provide users with more information, a preview of the full-text, and additional features such as an email function. Users bypassed these options and clicked on the title instead. The Discovery Usability Group theorized that users clicked on the title of the item because that behavior would be successful in a more familiar search interface like Google. The team customized the configuration so that a title click would open either the full-text of electronic items or the catalog record for physical items to accommodate users' instinctive search behaviors. The tabs, though a prominent feature of the discovery system, have proved to have little value for users.

Throughout the implementation of discovery systems in academic libraries, both research studies and anecdotal evidence have suggested that users do not find end-user features like facets valuable; however, discovery system vendors have made no apparent attempt to reimagine the possibilities for search refinements. Indeed, most of the findings in this study will present few surprises to anyone familiar with the discovery usability literature, which is itself concerning. As our literature review has shown, many of the same general usability issues have repeated throughout studies of Primo since 2008, and most are very similar to usability issues in other, competitor discovery systems. This raises some concerns about the pace of innovation in the discovery field, and whether discovery vendors are genuinely taking into account the research findings about the needs of our users as they refine their products. In a recent article, David Nelson and Linda Turney identified many issues with discovery facets in their current form that may be barriers to usage, particularly labeling and library jargon; we join them in urging vendors and libraries to collaborate more closely for deep analysis of actual facet usage by users, and to address those factors that have negatively affected facets' value.25

During our usability study, a common barrier to the successful completion of a task was not the technology itself but a lack of understanding of the task. Participants had difficulty deciphering a citation, which may have led to their tendency to search for a journal article by author and not by title. Many participants struggled with using call numbers, and how to find and check out books in the library. Peer review also proved to be a difficult or unfamiliar concept for many; when looking for peer-reviewed articles, some participants clicked on the "Reviews" facet, which limited their searches to an inappropriate resource type. Additionally, participants did not differentiate between journal articles and newspaper articles, which may indicate a broader inability to differentiate between scholarly and nonscholarly resources. This effect may be exaggerated by the high percentage of science students who participated, as these students may not have frequent need for newspaper articles. All of these challenges, however, are indicative of a deeper problem with terminology. Regardless of how simple it is to limit a search to peer-reviewed articles, a user who does not understand what peer review means cannot complete the task with confidence or certainty.

Librarians struggle with presenting understandable language and avoiding library terminology; as we discovered, academic language, like "peer-reviewed" and "citation," presents a similar problem. These are not issues that can be resolved with a technological solution. Rather, we join previous authors in suggesting that instruction may be a reasonable way to address many usability issues in Primo. From our findings and from those in the wider literature, we conclude that general instruction in information literacy is prerequisite for effective use of this or any research tool, particularly for undergraduates. Nichols et al. "recommend studying how to effectively provide instruction on Primo searching and results interpretation,"26 but instruction on the use of a single tool is of limited utility to students in their academic lives. Instead libraries could bolster information literacy instruction on key concepts around the production and storage of information, scholarly communications, and differences in information types. Teaching these concepts effectively should help to alleviate the most common user issues, including understanding terminology and different types of information, as well as helping students to understand key elements of research in general. This is a particularly important point to note for librarians working as advocates for information literacy instruction, especially in cases where administrators or faculty may feel that more advanced tools, like discovery systems, should make instruction obsolete.

8. CONCLUSION

Several changes were made to the Primo interface in response to breakdowns identified during the usability study. Resource Discovery Systems first implemented the changes to the Primo development sandbox. After the Discovery Usability and Advisory Groups agreed on the changes, they were made available on the live site (see figure 4). The redesigned search results page became available to the general public between the spring and summer academic sessions of 2015. In addition to the changes that were made because the usability study, RDS made changes to the look and feel to make the search results interface more aesthetically pleasing and more in line with the University of Houston brand.

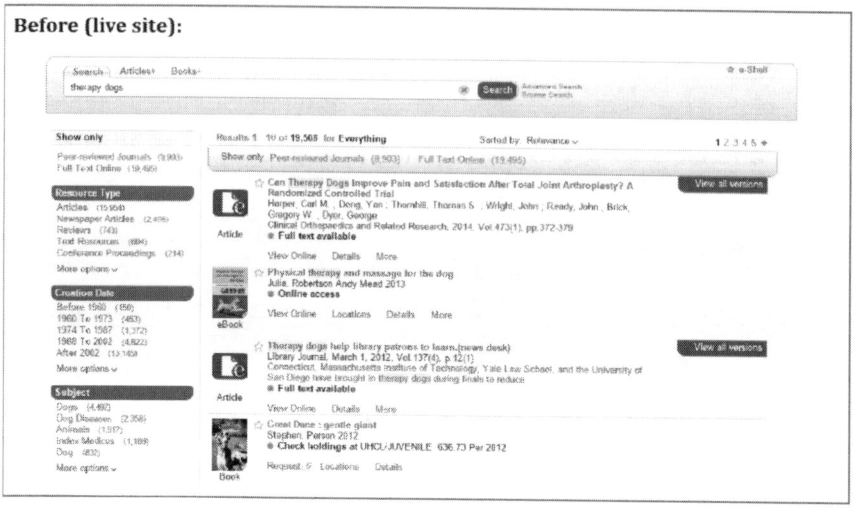

Figure 4. Primo Interface before Usability Testing

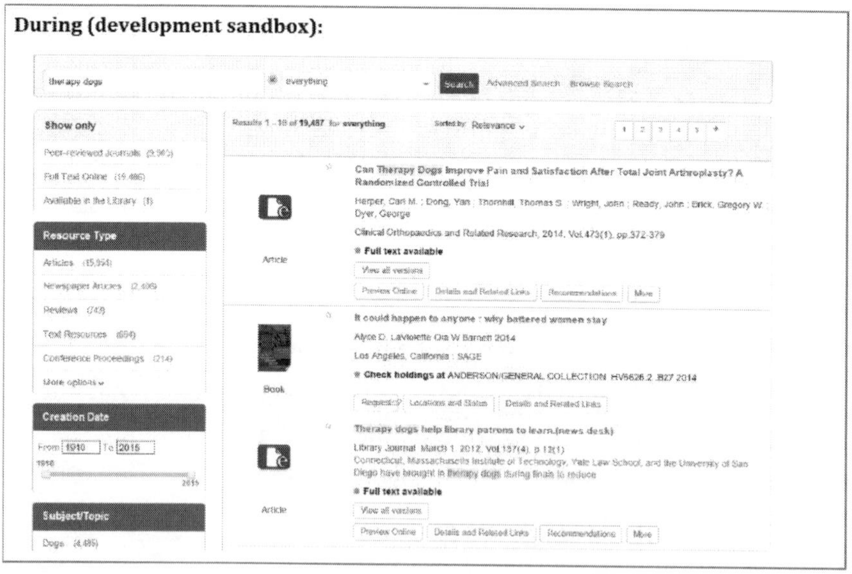

Figure 5. Primo Interface during Usability Testing

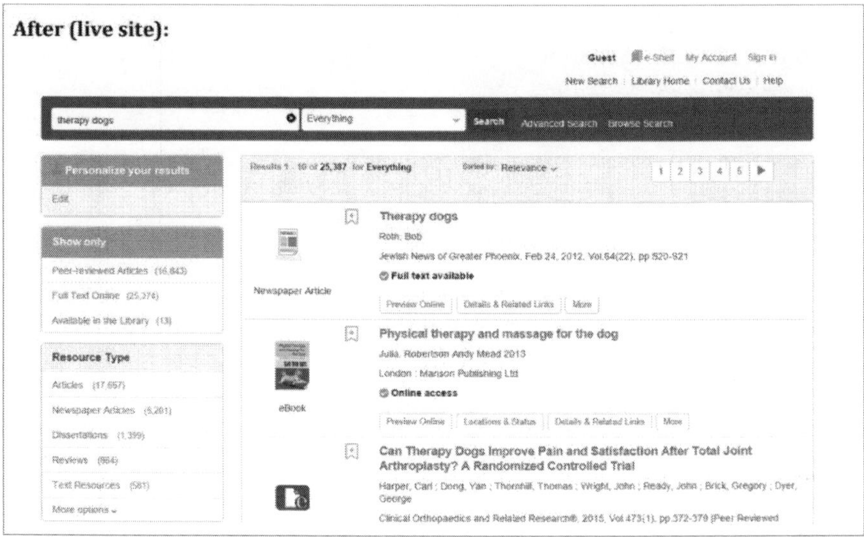

Figure 6. Primo Interface after Usability Testing

Many larger assertions of this study, encompassing implications for instruction and our needs from discovery vendors, will require further study to address. The authors intend to continue to investigate these issues as additional usability testing is conducted and to use the data to support future vendor relations and instructional curriculum development discussions.

REFERENCES

1. Tamar Sadeh, "User Experience in the Library: A Case Study," New Library World 109, no. 1/2 (2008): 7–24,

2. Aaron Nichols et al., "Kicking the Tires: A Usability Study of the Primo Discovery Tool," Journal of Web Librarianship 8, no. 2 (2014): 172–95, Scott Hanrath and Miloche Kottman, "Use and Usability of a Discovery Tool in an Academic Library," Journal of Web Librarianship 9, no. 1 (2015): 1–21,

3. David J. Comeaux, “Usability Testing of a Web-Scale Discovery System at an Academic Library,” College & Undergraduate Libraries 19, no. 2–4 (2012): 189–206

4. Kylie Jarrett, “FindIt@ Flinders: User Experiences of the Primo Discovery Search Solution,” Australian Academic & Research Libraries 43, no. 4 (2012): 278–300; Nichols et al., "Kicking the Tires." Sadeh, “User Experience in the Library.”

5. Jarrett, “FindIt@ Flinders”; Nichols et al., “Kicking the Tires.”

6. Xi Niu, Tao Zhang, and Hsin-liang Chen, “Study of User Search Activities with Two Discovery Tools at an Academic Library,” Libraries Faculty and Staff Scholarship and Research 30, no. 5 (2014), doi:10.1080/10447318.2013.873281; Hanrath and Kottman, “Use and Usability of a Discovery Tool in an Academic Library.”

7. Rice Majors, “Comparative User Experiences of Next-Generation Catalogue Interfaces,” Library Trends 61, no. 1 (2012): 186–207,; Niu, Zhang, and Chen, “Study of User Search Activities with Two Discovery Tools at an Academic Library.”

8. Beth Thomsett-Scott and Patricia E. Reese, “Academic Libraries and Discovery Tools: A Survey of the Literature,” College & Undergraduate Libraries 19, no. 2–4 (2012): 123–43, Sadeh, “User Experience in the Library.”

9. Comeaux, “Usability Testing of a Web-Scale Discovery System at an Academic Library.”

10. Jessica Mahoney and Susan Leach-Murray, “Implementation of a Discovery Layer: The Franklin College Experience,” College & Undergraduate Libraries 19, no. 2–4 (2012): 327–43, doi:10.1080/10691316.2012.693435.

11. Joy Marie Perrin et al., “Usability Testing for Greater Impact: A Primo Case Study,” Information Technology & Libraries 33, no. 4 (2014): 57–67.

12. Majors, "Comparative User Experiences of Next-Generation Catalogue Interfaces"; ThomsettScott and Reese, "Academic Libraries and Discovery Tools."

13. Jarrett, "FindIt@ Flinders"; Mahoney and Leach-Murray, "Implementation of a Discovery Layer."

14. Jarrett, "FindIt@ Flinders"; Mahoney and Leach-Murray, "Implementation of a Discovery Layer"; Nichols et al., "Kicking the Tires."

15. Jarrett, "FindIt@ Flinders"; Mahoney and Leach-Murray, "Implementation of a Discovery Layer"; Perrin et al., "Usability Testing for Greater Impact : A Primo Case Study."

16. Jarrett, "FindIt@ Flinders"; Nichols et al., "Kicking the Tires"; Hanrath and Kottman, "Use and Usability of a Discovery Tool in an Academic Library"; Majors, "Comparative User Experiences of Next-Generation Catalogue Interfaces."

17. Comeaux, "Usability Testing of a Web-Scale Discovery System at an Academic Library"; Thomsett-Scott and Reese, "Academic Libraries and Discovery Tools."

18. Jarrett, "FindIt@ Flinders."

19. Mahoney and Leach-Murray, "Implementation of a Discovery Layer"; Perrin et al., "Usability Testing for Greater Impact. "

20. Mahoney and Leach-Murray, "Implementation of a Discovery Layer"; Nichols et al., "Kicking the Tires"; Niu, Zhang, and Chen, "Study of User Search Activities with Two Discovery Tools at an Academic Library."

21. Thomsett-Scott and Reese, "Academic Libraries and Discovery Tools."

22. Tao Zhang and Merlen Prommann, "Applying Hierarchical Task Analysis Method to Discovery Layer Evaluation," Information Technology & Libraries 34, no. 1 (2015): 77–105,

23. David Nelson and Linda Turney, "What's in a Word? Rethinking Facet Headings in a Discovery Service," Information Technology & Libraries 34, no. 2 (2015): 76–91,

24. Nichols et al., "Kicking the Tires," 184.

Chapter 7

Fulfill Your Digital Preservation Goals with a Budget Studio

Yongli Zhou

Digital Repositories Librarian, Colorado State University Libraries, Fort Collins, Colorado.

ABSTRACT

To fulfill digital preservation goals, many institutions use high-end scanners for in-house scanning of historical print and oversize materials. However, high-end scanner prices do not fit in many small institutions' budgets. As digital single-lens reflex (DSLR) camera technologies advance and camera prices drop quickly, a budget photography studio can help to achieve institutions' preservation goals. This paper compares images delivered by a high-end overhead scanner and a consumer-level DSLR camera, discusses pros and cons of using each method, demonstrates how to set up a cost-efficient shooting studio, and presents a budget estimate for a studio.

1. INTRODUCTION

Colorado State University Libraries (CSUL) are regularly engaged in a variety of digitization projects. Materials for some projects are digitized in-house, while items from selected projects are sometimes outsourced. Most fragile materials that require professional handling are digitized inhouse using an expensive overhead scanner. However, the overhead scanner has been occasionally unstable since it was purchased, and this has delayed some of our digitization projects. As digital photography technologies advance, image quality delivered by digital singlelens reflex (DSLR) cameras is improving, and camera prices have lowered to an affordable level. In this paper, I will compare images produced by a scanner and a camera side-by-side, list pros and cons of using each method, illustrate how to establish a shooting studio, and present a budget estimate for that studio.

2. LITERATURE REVIEW

There are many online guidelines and manuals for digitizing print materials. Some universities and museums have information about their digitization equipment online. Most articles focus on either high-end scanners or customized scanning stations. These articles are very helpful for universities and museums that are relatively well funded. However, there is almost no literature discussing how to use inexpensive digital cameras and photography equipment to produce highquality digitized images. This article will use a case study to prove that a low-budget studio can produce high-quality digitized images.

2.1 Comparison of Scanned and Photographed Images

The test camera set was chosen because it was the one the author used for general purpose. The camera was also chosen by many professional photographers because of its quality and affordability. To avoid dispute, the overhead scanner's make and model are not revealed.

Table 1. Test Equipment

Budget Studio	Overhead Scanner
• Nikon D800 • Nikon AF Micro-Nikkor 60mm f/2.8D Lens • Manfrotto 055CXPRO3 3-Section Carbon Fiber Tripod Legs • Really Right Stuff BH-40 LR II Ballhead • Nonreflective glass • Book cradles • X-Rite Original ColorChecker Card • Natural daylight • Total cost: **$4,500** and no maintenance fees (priced in 2014)	• Our overhead scanner • Nonreflective glass • Book cradles • Purchase price: **$55,000** (purchase in 2007) • **$8,000 annual maintenance** (2013 price)

2.2 Focus and Sharpness

A quality digitized image needs to have a good focus. A well-focused image shows details better and can produce better Optical Character Recognition (OCR) results for text-based documents. At CSUL, we have no control over the automatic focus on our overhead scanner and have noticed that sometimes one page is sharply focused but the next page is slightly out-of-focus. During the scanning process, our overhead scanner does not indicate if a shot is focused or not. A DSLR camera can beep or display a flashing dot on the viewfinder when in focus.

2.3 Illustration

The following two figures compare images produced by our test DSLR and overhead scanner. Both images were originals and have not been enhanced by software. In addition to this image, we tested nine other illustrations. Following our comparison study, we concluded that a semiprofessional DSLR camera produces sharper images than our expensive overhead scanner.

In figure 1, at 100 percent zoom, the left image has a better focus, contains more details, and has colors closer to the original. The left image was taken using a Nikon D800 + Nikkor 60mm macro lens and under natural lighting. The right image was produced by our overhead scanner.

In Figure 2, at 200 percent zoom, the left image (taking using the DSLR) shows much more detail than the image on the right (taken with the overhead scanner).

Figure 1. Comparative Images from DSLR (Left) and Overhead Scanner (Right), at 100 Percent Zoom. Image from Samuel M. Janney, The Life of William Penn; with Selections from His Correspondence and Auto-Biography (Philadelphia: Hogan Perkins & CO, 1852), plate between pages 296 and 297.

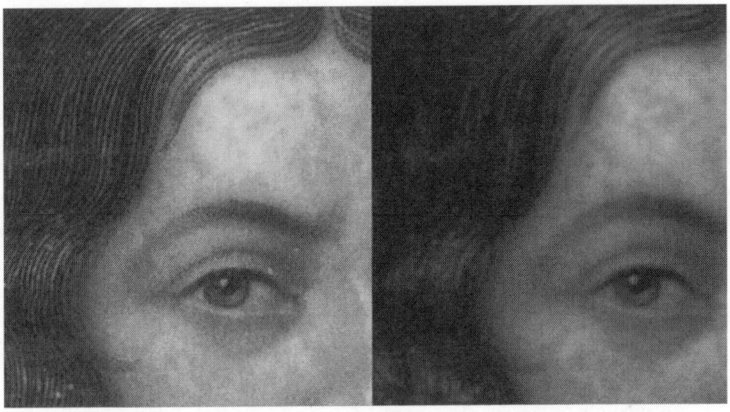

Figure 2. Comparative Images from DSLR (Left) and Overhead Scanner (Right), at 200 Percent Zoom. Image from Samuel M. Janney, The Life of William Penn; with Selections from His

Correspondence and Auto-Biography (Philadelphia: Hogan Perkins & CO, 1852), frontispiece, print.

At CSUL, the process of digitizing a text document includes scanning pages, converting them into Portable Document Format (PDF) files, and applying an OCR process. In general, a well-focused image of text produces better OCR results, although software such as Adobe Acrobat can tolerate fuzzy images and produce reasonably accurate OCR text. Our OCR tests from a slightly out-of-focus image and a well-focused image have no significant difference; however, from preservation and usability standpoints, we prefer well-focused images.

2.4 Photograph

A 6.5 inches by 4.5 inches silver print was used for this test. Our tests show that the test DSLR camera produced a sharper image of this historic photograph.

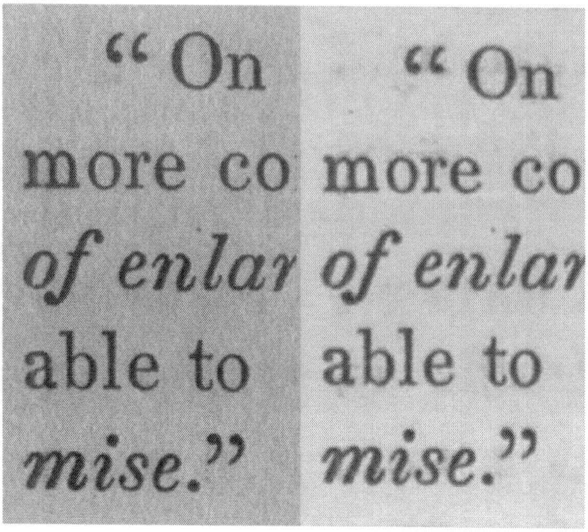

Figure 3. The left image was produced by our test DSLR camera and has a better focus. The right image was produced by our overhead scanner. Samuel M. Janney, The Life of William Penn; with Selections from His Correspondence and Auto-Biography (Philadelphia: Hogan Perkins & CO, 1852), 300, print.

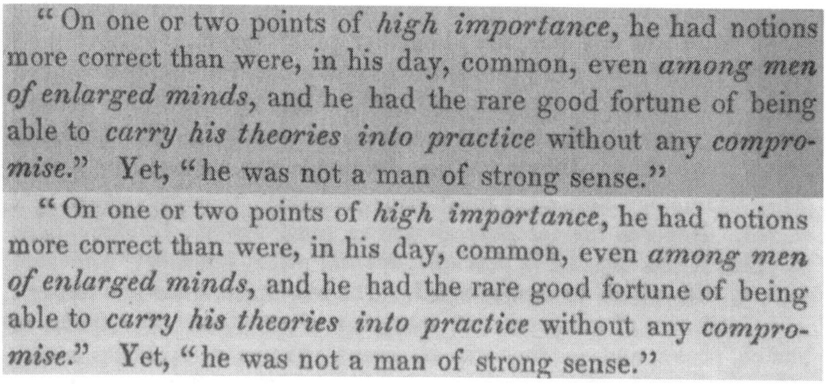
"On one or two points of *high importance*, he had notions more correct than were, in his day, common, even *among men of enlarged minds*, and he had the rare good fortune of being able to *carry his theories into practice* without any *compromise*." Yet, "he was not a man of strong sense."

"On one or two points of *high importance*, he had notions more correct than were, in his day, common, even *among men of enlarged minds*, and he had the rare good fortune of being able to *carry his theories into practice* without any *compromise*." Yet, "he was not a man of strong sense."

Figure 4. We ran the OCR process on the above two images. The top image was produced by our test DSLR camera and the bottom image was produced by our overhead scanner. Samuel M. Janney, The Life of William Penn; with Selections from His Correspondence and Auto-Biography (Philadelphia: Hogan Perkins & CO, 1852), 300, print

Table 2. OCR Results Comparison

Generated from the Image by Camera	Generated from the Image by Scanner
" On one or two points of *high importance,* he had notions more correct than were, in his day, common, even *among men of* **e1~larged** *minds,* and he had the rare good fortune of being able to *carry his theories into practice* without any *compromise."* Yet, "he was not a man of **stron** sense."	" On one or two points of *high importance,* he **bad** notions more correct than were, in his day, common, even ***arnong*** *men* ***of*** **e1~larged** *minds,* and he had the rare good fortune of being able to *carry his theories into practice* without any *compromise."* Yet, "he was not a man of strong sense."

These test results are very close because of the forgiveness of the Adobe Acrobat software. However, we have seen that for some other pages, a better-focused image generates improved OCR results.

Figure 5. Tested 6.5 Inches by 4.5 Inches Photograph. The red square indicates the enlarged area for figure 6. Historical photograph from Colorado State University Archives and Special Collections.

Figure 6. Screen View at 100 Percent Zoom of a Silver Print. The top image was produced by the test DSLR camera and the bottom one was produced by our overhead scanner. Historical photograph from Colorado State University Archives and Special Collections.

2.5 Oversize Materials

For oversized materials, overhead scanners and DSLR cameras have their drawbacks, so we do not think either option is ideal for them. Our library uses a map scanner to scan oversize maps and posters. However, a map scanner is expensive and may not fit many libraries' budgets. A map scanner also is not suitable for fragile maps or posters.

Our overhead scanner's maximum scanning area is 24 inches by 17 inches, and the test map's size is 25 inches by 26 inches. We had to scan the map in four sections and stitch them together using Adobe Photoshop. Each section image has a files size of 313 MB. Because of large file sizes, the stitching process is extremely slow. Also stitching images is not recommended because there are always some degrees of mismatching errors created by lens distortion.

A camera can capture any material size, but the details of the photographed images diminish as the material's size increases. The photo of the entire map taken by our test DLSR has a file size of 35.8 MB. The image produced by camera has a lower resolution and less detail.

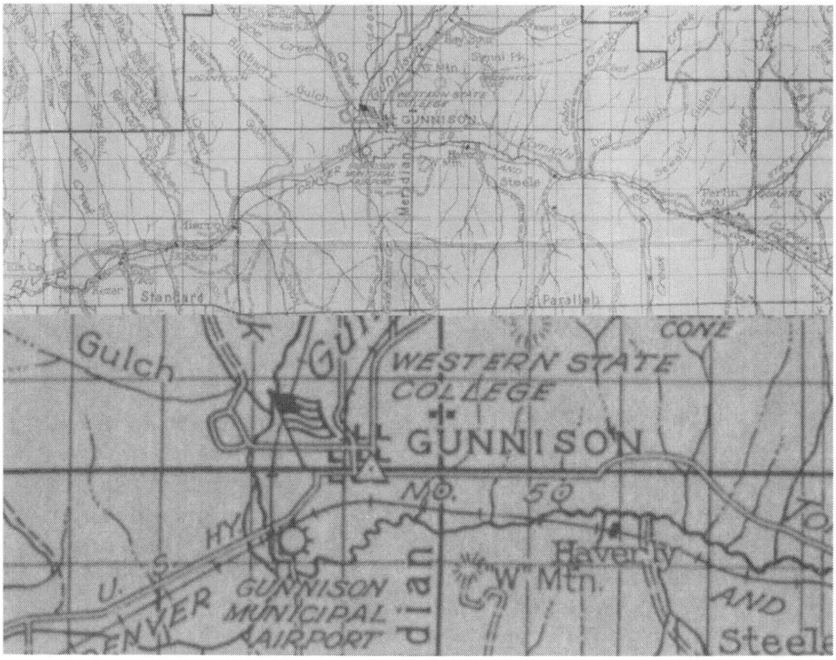

Figure 7. Oversized Materials Screen View at 100 Percent Zoom. The top image was photographed by the test DSLR. The bottom image was scanned by our overhead scanner. Historical map from Colorado State University Archives and Special Collections.

2.6 Small Prints

One big advantage of a DSLR camera is that it can be set farther away to take pictures of oversized materials or very close to smaller objects to take close-up pictures. Comparatively, the distance of lens and scanning platform on our overhead scanner is fixed, so no close-up images can be produced, and everything is reproduced at scale of 1:1.

For the following example, we used a 5.5 inches by 3.5 inches drawing as our test subject.

Figure 8. A 5.5 inches by 3.5 inches Fine Drawing. A historical booklet from Colorado State University Archives and Special Collections.

The image produced by our overhead scanner has a resolution of 3,427 pixels by 2,103 pixels. The camera produces a 6,776 pixels by 4,240 pixels image. The higher pixel count allows users to see more details at the same zoom level. The image produced by camera is not only sharper but also contains more details. It also is good for making enlarged prints for promotion materials.

For smaller maps, a DSLR camera also produces superior images. For the following sample, we tested a 15 inches by 9.5 inches map.

Figure 9. Small Prints Screen View at 100 Percent Zoom. The left image is produced by a DSLR with a macro lens and the right image was scanned by our overhead scanner. A historical booklet from Colorado State University Archives and Special Collections.

2.7 Post-Processing

2.7.1 Use of a Sharpening Filter

Our tests showed that a main drawback of our overhead scanner is that images produced are outof-focus. Some digitization guidelines recommend minor post-processing for delivered images files to improve image quality. One might argue that to fix our overhead scanner's out-of-focus problem, sharpening can be applied. Technical Guidelines for Digitizing Cultural Heritage Materials: Creation of Raster Image Master Files recommends doing minor post-scan adjustment to optimize image quality and bring all images to a common rendition.1 This is good advice, but it is not applicable in real-world practice. To get the best result, each image would need to be evaluated and have a sharpening filter applied separately because when an improper sharpening setting is applied to an image, it often creates haloing artifacts and an unnatural look. The application of a sharpening filter to each image process will be extremely time-consuming.

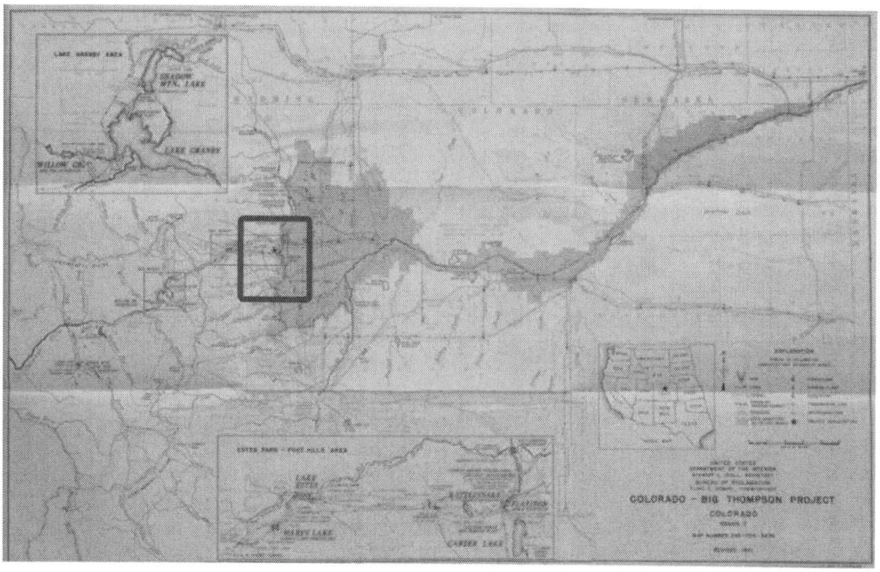

Figure 10. A 15 inches by 9.5 inches map. The blue square indicates the enlarged area for figure 11. Historical map from Colorado State University Archives and Special Collections.

The haloing artifact is also called chromatic aberration (CA) effect. CA appears as unsightly color fringes near high contrast edges. Chromatic aberrations are typically only visible when viewing the image on-screen at higher zoom levels or on large prints.

The following example shows that the CA may not appear at lower zoom levels, such as 50 percent or 100 percent. The left image has no sharpening filter applied and the right image has a sharpening filter applied. At 100 percent zoom, chromatic aberration is almost not identifiable, and the right image appears to be superior in turns of sharpness.

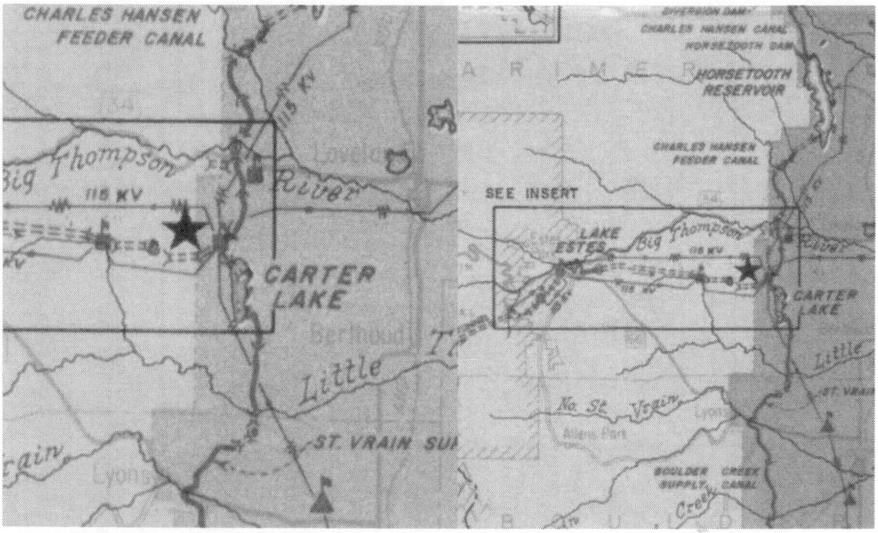

Figure 11. Small Map Screen Views at 100 Percent Zoom. The left image was photographed by a DSLR camera with a macro lens and the right image was produced by our overhead scanner. Historical map from Colorado State University Archives and Special Collections.

At a higher zoom level, we see CA, visible in the right image of figure 13. The extra colors are introduced by the software.

We recommend not applying sharpening filters to original scanned images; instead, attempt to obtain well-focused images from the beginning. For this reason, the test DSLR camera outperformed our overhead scanner for most materials.

Figure 12. Sharpening Filter Comparison Sample at 100 Percent Zoom. The left image has no sharpening filter applied and the right image has been applied a sharpening filter. Historical map from Colorado State University Archives and Special Collections.

Figure 13. Comparison of Sharpening Filter Applied to Images and at 500 Percent Zoom. The left image has no sharpening filter applied and the right image has sharpening filter applied. Historical map from Colorado State University Archives and Special Collections.

2.7.2 Color Balance

Have you seen a scanned color image or color photograph with colors very different from the original image? For example, a white area appears to be bluish, or it has an orange cast? When scanning or photographing an image under different lighting, the output image can have very different colors.

In the following figure, the left image was shot at a correct white balance (WB) setting. WB is the process of removing unrealistic color casts so that objects that appear white in person are rendered white in your photo.2 The center image has a blue color cast, which was caused by a lower Kelvin setting, and the right image was shot at a higher Kelvin setting. A camera may create images with the wrong colors, but so will a scanner if it is not calibrated correctly.

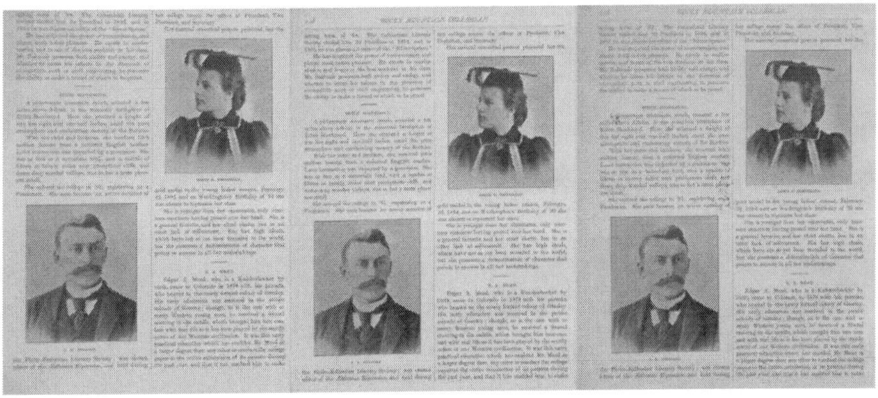

Figure 14. Images Shot under Different White Balance Settings.

We pay an $8,000 annual service fee for overhead scanner maintenance, which includes scanner color calibration. In general, image colors rendered by the machine are close to original colors but not exact. We have noticed that some images have a very light green overcast and other others are overly yellow; sometimes images appear to be darker than they should be. Because we are not certified to calibrate the overhead scanner, we only use the prescribed settings set by technicians. Also, we have no control over maintaining a fading light bulb, which will affect correct exposure.

WB adjustment on photographs taken in a studio can be very precise. Most DSLR contains a variety of preset white balances. In general, auto WB works well, but does not deliver the best results. Custom WB allows fine-tuning of colors. If a shooting studio is set up properly, the lighting should be consistent, so ideally one setting found most desirable can be used repeatedly. However, professional photographers do test shots at the beginning of each shooting session. Once they find the optimal test shot, they will use the exact settings for the batch. Later, they will do minor color adjustment on the chosen test shot to ensure precise color representation, and then apply the adjustment settings on all other photos of the same batch. Because many small variations can be present for each shooting session, they do not use the settings from the previous shooting. It may seem arduous to do test shots for each shooting, but it ensures accurate color reproduction.

Many professional photographers use ColorChecker Passport, 3 which is a commercial product to help with quick and easy capture of accurate colors. I will demonstrate briefly a useful trick I learned from a professional photography seminar how to utilize ColorChecker Passport to apply correct white balance a group of images. [4]

Step 1: Place an 18 percent gray card or a ColorChecker Passport card on top of a page. Choose the correct exposure and take the photo. Use the same exposure setting to take additional photos. For demonstration purposes, we deliberately used a very low and high Kelvin setting for sample images. The low Kelvin setting created cool and blue tones and the high Kelvin setting created a tone that was too warm. Note that the test shot with ColorChecker Board was not taken with exactly the correct white balance setting.

Figure 15. Sample Images for White Balance Adjustment. Rocky Mountain Collegian 3–4 (1893), 118, Colorado State University Archives and Special Collections.

Step 2: In Adobe Lightroom, select the test target image and switch to "Develop" mode. Select the White Balance tool, move the cursor over a gray area, try to find a spot where the red, green, and blue (RGB) values are close. If you can find a place with equal RGB values, it will be ideal. This simple click will set the test image's white balance to an almost perfect setting.

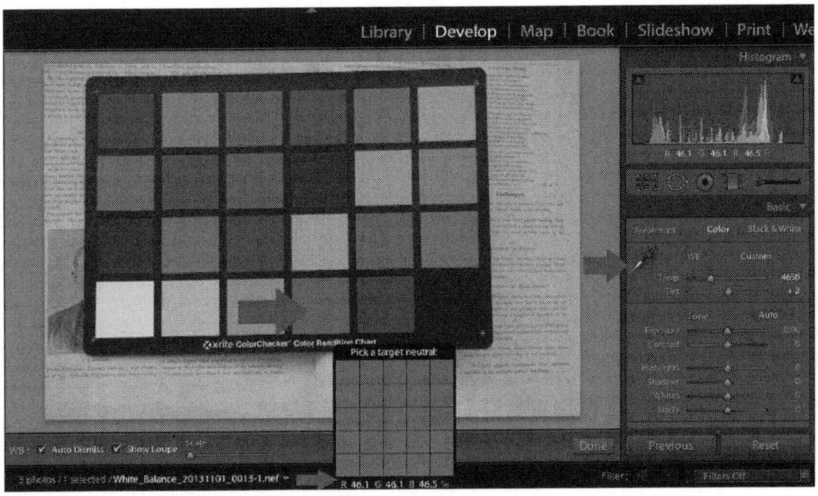

Figure 16. Applying a White Balance in Adobe Lightroom 4

Step 3. Synchronize other images' settings with the target image. Select the target image and all other images, click the Sync button, and select settings you would like to synchronize. Make sure the WB button is checked.

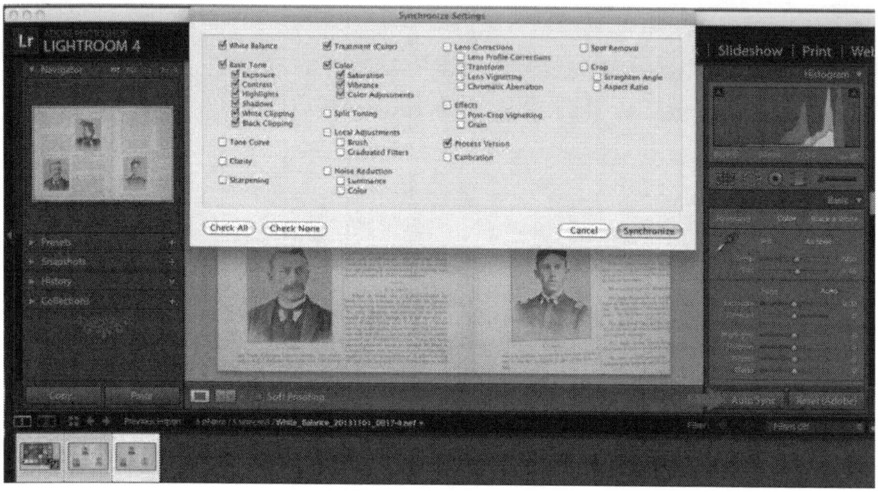

Figure 17. Synchronize Settings in Adobe Lightroom 4

Figure 18. Synchronized Images with Correct White Balance. Rocky Mountain Collegian 3–4 (1893), 118, Colorado State University Archives and Special Collections.

Recently, I had the opportunity to visit the Spencer Museum of Art's digitization lab. They have a different workflow to ensure even more scientifically correct colors. If you are interested in their approach, you can contact their information technology manager or photographer.

2.7.3 Color Space

One very important thing to understand is color space when you use a DSLR camera. Many DSLR cameras support Adobe RGB and sRGB.

sRGB reflects the characteristics of the average cathode ray tube (CRT) display. This standard space is endorsed by many hardware and software manufacturers, and it is becoming the default color space for many scanners, low-end printers, and software applications. It is the ideal space for web work but not recommended for prepress work because of its limited color gamut.

Adobe RGB (1998) was designed to encompass most of the colors achievable on CMYK printers, but only by using RGB primary colors on a device such as your computer display.5 It is recommended to use this color space if you need to do print production work with a broad range of colors. Many scanning vendors deliver images in Adobe RGB color space.

ProPhoto RGB contains all colors that are in Adobe RGB, and Adobe RGB contains nearly every color that is in sRGB. This color space covers more colors than the human eye can see. It can only be used for images in RAW format and in 16-bit mode. Common file formats that support 16-bit images are TIFF and PSD. Most printers do not support 16-bit format. This color space normally is used by photographers who have a specific workflow and who print on specific high-end inkjet printers. When converting from 16-bit to 8-bit, some images will have banding or posterization problems.

Banding is a digital imaging artifact. A picture with banding problem shows horizontal or vertical lines.

Figure 19. An Example of Colour Banding, Visible in the Sky in This Thotograph.[6]

Posterization of an image entails conversion of a continuous gradation of tone to several regions of fewer tones, with abrupt changes from one tone to another.[7]

Figure 20. An Example of Posterization.8

While it is a good idea to capture images using Adobe RGB to preserve a wide range of colors, you should convert images to sRGB when delivering to unknown users and displaying on the web. Currently, sRGB is the only appropriate choice for images uploaded to the web, since most web browsers don't support any color management. Adobe RGB images that are uploaded to websites without conversion to sRGB generally appear dark and muted.9 If they were printed on printers that do not support Adobe RGB format, colors will be dull too.

3. SETTING UP A BUDGET STUDIO

3.1 Commercial Approach

BookDrive Pro is a commercially available digitization unit. It uses two digital cameras and built-in flash lights. It may be the optimal solution for your projects, but it also may not fit your library's budget. The unit also is not suitable for oversized material such as large maps and posters. For more information about this product, please visit http://pro.atiz.com/.

3.2 Sample Budget Studio Setup

A digitization lab can have three rooms or areas, one for oversized materials, one for smaller prints or 3-D objects, and one for computers.

The area for shooting oversized materials should have black walls and floor. You can either use one flash light to bounce light off the ceiling or use two flash lights to shine lights directly onto the materials. For fragile materials, the first approach is more appropriate.

The area for shooting smaller prints or 3-D objects should have a stable table and black or white background paper. For this room or area, black walls and floor are not required.

For shooting equipment, I will use the set chosen by the photographer from the University of Kansas Spencer Museum of Art as my example.

The total cost for a "budget" shooting studio ranges from $10,000 to $15,000, and there is no annual maintenance expense.

Functions of Some Elements in the Sample Shooting Studio

1) Macro Lens: It allows close up shooting of objects. It is especially useful when photograph small prints and small 3-D objects. It can also be used to photograph regular and oversized materials.

2) Heavy-duty mono stand: It replaces a traditional tripod. It is very stable and allows quick adjustment of camera height and location.

3) Strobe, power pack, and reflector: Together they generate consistent and homogeneous light distribution. Recommended further reading: "Introduction to OffCamera Flash: Three Main Choices in Strobe Lighting."10

4) Light stand: It holds strobe and reflector.

5) Light meter: Hand-held exposure meters measure light falling onto a light-sensitive cell and converts it into a reading that enables the correct shutter speed and or lens aperture settings to be made.11

6) Book cradles: They help to minimize the stress on bookbindings and minimize page curvature problem.

7) Nonreflective glass: It helps to flatten a photographed page and reduce the reflection. However, it does not completely eliminate glass reflection. One very useful trick to reduce glass reflection is to place a black board with a hole above a page and shoot through the hole. This approach actually does not eliminate reflection but reflects black to the photograph. When the photograph is reviewed on computer, it will appear as no reflection has occurred.

Table 3. List of Items Needed to Prepare for a Budget Studio

Item Name	Sample Item	Purchasing URL	Price
DSLR camera	Nikon D810	http://www.bhphotovideo.com/c/search?atclk=Camera+Model_Nikon+D810&ci=6222&N=4288586280+3907353607	$2,996.95
Macro lens	Nikon AF Micro-Nikkor 60mm f/2.8D Lens	http://www.bhphotovideo.com/c/product/66987-GREY/Nikon_1987_AF_Micro_Nikkor_60mm_f_2_8D.html	$429.00
Heavy duty mono stand	Arkay 6JRCW Mono Stand Jr with Counter Weight—6'	http://www.bhphotovideo.com/c/product/2727-REG/Arkay_605138_6JRCW_Mono_Stand_Jr.html	$678.50
Strobe	Broncolor G2 Pulso—1600 Watt/Second Focusing Lamphead with 16' Cord	http://www.bhphotovideo.com/c/product/259745-REG/Broncolor_32_115_07_G2_Pulso_with_16.html	$3,053.68
Power pack	Broncolor Senso A4 2,400W/s Power Pack	http://www.bhphotovideo.com/c/product/745060-REG/Broncolor_31_051_07_Senso_A4_2_400W_s_Power.html	$3,629.92
Reflector	Broncolor P65 Reflector, 65 Degrees, 11" Diameter, for Broncolor Pulso 8, Twin and HMI	http://www.bhphotovideo.com/c/product/7162-REG/Broncolor_33_106_00_P65_Reflector_65_Degrees.html	$513.52
Reflector	Broncolor Softlight Reflector, 20" Diameter, for Broncolor Primo, Pulso 2/4 & HMI Heads	http://www.bhphotovideo.com/c/product/7167-REG/Broncolor_33_110_00_Softlight_Reflector_20_for.html	$501.76
Light Stand	Impact Air-Cushioned Light Stand	http://www.bhphotovideo.com/c/product/253067-REG/Impact_LS10AB_Air_Cushioned_Light_Stand.html	$44.99
Light meter	Sekonic L-308S Flashmate—Digital Incident, Reflected and Flash Light Meter	http://www.bhphotovideo.com/c/product/368226-REG/Sekonic_401_309_L_308S_Flashmate_Light_Meter.html	$199.00
Book cradle	Book Exhibition Cradles	http://www.universityproducts.com/cart.php?m=product_list&c=1115&primary=1&parentId=1271&navTree[]=1115	$30.00
Background paper	Savage Seamless Background Paper (Both white and black)	http://www.bhphotovideo.com/c/product/45468-REG/Savage_1_12_107_x_12yds_Background.html	$45.00 x 2 = $90.00
Nonreflective glass	1/4" Optiwhite Starphire Purified Tempered Single Lite Clear Class	Can be purchased at local glass store.	$75.00
White balancing accessory	X-Rite Original ColorChecker Card	http://www.bhphotovideo.com/c/product/465286-REG/X_Rite_MSCCC_Original_ColorChecker_Card.html	$69.00
Software	Adobe Lightroom 5	http://www.adobe.com/products/photoshop-lightroom.html	$150.00

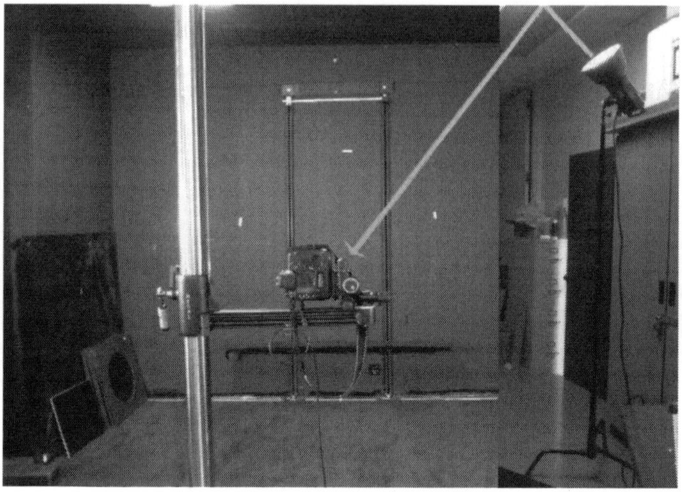

Figure 21. The University of Kansas Spencer Museum of Art Digitization Lab Setup for Oversized Materials

Many librarians believe that digitizing print materials using a digital camera requires a professional photographer, but this is not necessarily true. A professional photographer or even an art student can act as a consultant to help set up a shooting studio and provide basic training. Also, many museums have professional photographers and have set up shooting studios for digitization. They are very willing to share their experience and even provide training. I believe the learning curve for operating a shooting studio is no greater than the learning curve to operate an overhead scanner machine and its software.

3.3 Pros and Cons

No digitization equipment or system is perfect. They all have trade-offs in image quality, speed, convenience of use, quality of accompanying software, and cost. Our tests show that for most archival materials a DSLR camera will do a better job than an overhead scanner.

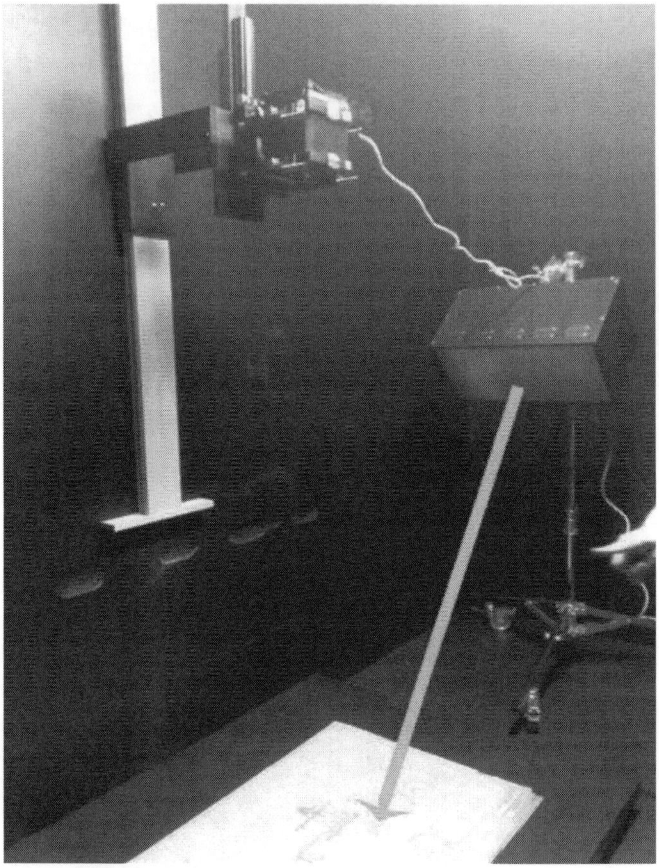

Figure 22. Steelworks Museum of Industry and Culture's Digitization Lab Setup for Oversized Materials

Pros of Overhead Scanner

- The scanner is a complete scanning station. It can be connected to a computer and starts scanning immediately. Materials can be placed on the scanning surface, so no equipment adjustments are required while scanning.

- It can scan and save images in bitmap format directly, while a DSLR camera can only shoot in grayscale or color.

- Built-in book cradles help to scan thick books and those that cannot be fully opened.
- Book curve correction functionality is provided by the accompanying software.

Figure 23. The University of Kansas Spencer Museum of Art Digitization Lab Setup for Smaller Prints and 3-D Objects

Cons of Overhead Scanner

- High cost. The overhead scanner we have cost more than $50,000, with an annual maintenance contract of $8,000.
- High replacement cost. When a scanner is outdated or broken, the entire machine has to be replaced.

- Instability. Our overhead scanner is unstable even when placed on a sturdy table and handled only by professionals. From April 2010 to October 2010, the scanner was down for a total of forty-two working days (sixty calendar days). The company fixed the machine onsite many times, but it continues to have minor problems and has not been completely reliable.

- The autofocus feature does not work consistently.

- Special training is needed to operate the machine and associated software.

- File formats supported are limited. Most scanners only support TIFF, JPEG, JPEG 2000, Windows BMP, and PNG.

- Unsupported outdated software: Our overhead scanner's software can only be run on an older operating system (Windows XP) because there is no updated software for this model.

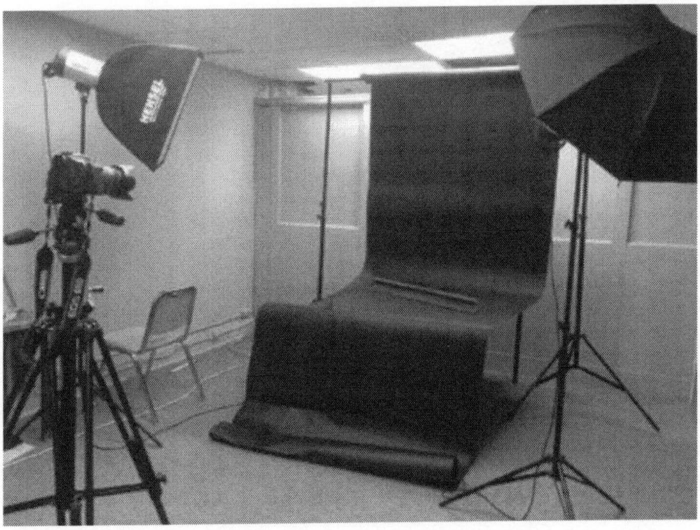

Figure 24. Steelworks Center of the West's Digitization Lab Setup for 3-D Objects

Figure 25. The University of Kansas Spencer Museum of Art Digitization Lab Setup for Materials Needed be Pressed Down by a Glass.

Pros of Budget Studio

- Stable. Under normal use DSLR cameras are much less likely to break down than scanners. For example, I have had an older DSLR, Nikon D200, for seven years. It has survived numerous backpacking trips, multiple drops, and extreme weather conditions. The camera still functions as needed.

- Fast and accurate focus. DSLR cameras are designed to focus quickly, and their focus indicators provide instant feedback to the operators so they know that the image is focused. If operated properly, images delivered by DSLR cameras can be sharper than ones delivered by scanners.

- Less expensive. A good quality DSLR camera and a lens can be purchased for fewer than $4,000 and last for years. As technologies advance, DSLR cameras' prices will continue to drop.

- Ability to save files in more formats. In addition to TIFF and JPEG formats, most DSLR cameras can save photos in RAW file format. Some cameras can directly save images in Digital Negative (DNG) format, and others deliver images in proprietary formats that can be converted into DNG using a computer program. Editing RAW images is nondestructive, while editing of TIFF and JPEG images is irreversible.

- Accurate WB and exposure. By using right shooting and post-processing techniques, photographs can have exact color reproduction. On the other hand, calibrating an overhead scanner most likely can only be performed by a company's trained technician. Proper exposure and WB are not guaranteed.

- The RAW file format usually provides more dynamic range. Overexposed and underexposed images can be fixed by adjusting exposure compensation via software; thus lost shadow or highlight detail can be restored.

- Can photograph 3-D objects. Archival collections often have materials other than books, such as art pieces. These materials are better to be photographed than scanned.

- Versatile. Cameras can perform on-site digitization, while overhead scanners are too bulky to be moved around.

- Faster and better preview. Images can be viewed instantly on a computer when proper software, such as Adobe Lightroom, is used. Operators can compare multiple shoots on a screen side-by-side and decide which photo to retain.

- More accessible technical support. The number of DSLR camera users is much higher than overhead scanner users. Technical questions can often be answered through online forums.

- Easy to find replacement parts. When a piece in a shooting studio break down, it is easy to find replacement piece and replace by staff.

- Easy software updates. Software used in a studio is independent from equipment.

Cons of Budget Studio

- There is learning curve for setting up a shooting studio, operating the studio, and mastering new image processing techniques.
- A DSLR camera with a lower pixel setting will not be sufficient for scanning large-format materials, such as posters and maps.
- No built-in book curve correction is provided by Adobe Photoshop or Lightroom. However, our experience proves that the automatic book curve function does not always work well. We normally use a home-made book cradle to help lay a page flat and use one or two weights to hold down the other side of book. For some books, if flatness is hard to achieve, we place a piece of glass on the top to ensure the flatness.
- Security concern: Since a DSLR camera is highly portable, it can be stolen easily.

4. CONCLUSION

The technology of DSLR cameras has advanced very quickly in the past ten years. Newer DSLR cameras can handle higher resolutions and have very little image noise even at a high ISO setting. The higher demand for DSLR cameras and accompanying image-editing software results in more rapid technology advances compared to low-demand and high-end overhead scanners. High consumer demand drives DSLR camera prices much lower than prices for overhead scanners. In addition, the wide range of consumers purchasing DSLR cameras and software prompts companies to offer more user-friendly interfaces. As you can see from our tests, for most library materials a DSLR camera can produce superior images. If you do not have a budget for high-end overhead scanners, you can still fulfill your digitization preservation goals with a budget studio.

Figure 26. Scanning Setup Using a Book Cradle.

ACKNOWLEDGEMENT

I would like to thank Robert Hickerson and Ryan Waggoner, the University of Kansas Spencer Museum of Art, Tim Hawkins, and Steelworks Center of the West for showing their digitization labs and sharing experience with me.

REFERENCES

1. Federal Agencies Digitization Guidelines Initiative, "Technical Guidelines for Digitizing Cultural Heritage Material: Creation of Raster Image Master Files," August 2010, http://www.digitizationguidelines.gov/guidelines/digitize-technical.html

2. "Tutorials: White Balance," Cambridge in Colour, accessed March 9, 2016, http://www.cambridg eincolour.com/ tutorials /white-balance.htm.

3. "ColorChecker Passport User Manual," X-Rite Incorporated, accessed March 9, 2016, http://www.xrite.com/ documents/manuals/en/ColorCheckerPassport_User_Manual_en.pdf.

4. Scott Kelby, "Scott Kelby's Editing Essentials: How to Develop Your Photos," Pearson Education, Peachpit, accessed March 9, 2016, http://www.peachpit.com/articles/article.aspx?p=2117243&seqNum=3

5. "sRGB vs. Adobe RGB 1998," Cambridge in Colour, accessed March 9, 2016, http://www.cambridgeincolour.com/tutorials/sRGB-AdobeRGB1998.htm.

6. "Colour Banding," Wikipedia, accessed March 9, 2016, http://en.wikipedia.org/wiki/Colour_banding.

7. "Posterization," Wikipedia, accessed March 9, 2016, http://en.wikipedia.org/ wiki/Posterization.

8. "Image Posterization," Cambridge in Colour, accessed March 9, 2016, http://www.cambridgeincolour .com/tutorials/ posterization.htm.

9. Richard Anderson and Peter Krogh, "Color Space and Color Profiles," American Society of Media Photographers, accessed March 9, 2016, http://dpbestflow.org/color/color-space-andcolor-profiles.

10. Tony Roslund, "Introduction to Off-Camera Flash: Three Main Choices in Strobe Lighting," Fstoppers (blog), accessed March 9,

2016, https://fstoppers. com/originals/introductioncamera-flash-three-main-choices-strobe-lighting-40364.

11. "Introduction to Light Meters," B & H Foto & Electronics Corp., accessed March 9, 2016, http://www.bhphotovideo. com/find/ Product_Resources /lightmeters1.jsp.

Chapter 8

Transitioning from XML to RDF: Considerations for an Effective Move Towards Linked Data and the Semantic Web

Juliet L. Hardesty

Metadata Analyst at Indiana University Libraries, Bloomington, Indiana.

1. INTRODUCTION

Metadata, particularly within the academic library setting, is often expressed in eXtensible Markup Language (XML) and managed with XML tools, technologies, and workflows. Software tools such as the Oxygen XML Editor and querying languages such as XPath and XQuery over time have become capable of helping that management. However, managing a library's metadata currently takes on a greater level of complexity as libraries are increasingly adopting the Resource Description Framework (RDF). Semantic Web initiatives are surfacing in the library context with experiments in publishing metadata as Linked

Data sets, BIBFRAME development using RDF, and software developments such as the Fedora 4 digital repository using RDF.

Challenges are evident when considering examples of transitions from XML into RDF and show the need for communication and coordination between efforts to incorporate and implement RDF. This article outlines these challenges using different use cases from the literature and first-hand experience. The follow-up discussion considers ways to progress forward from metadata formatted in XML to metadata expressed in RDF. The options explored are not only targeted to metadata practitioners considering this transition but also to programmers, librarians, and managers.

2. LITERATURE REVIEW AND CONCEPTS

As an initial example of the challenges faced when considering RDF, clarifying terminology is still a helpful activity. RDF focuses on sets of statements describing relationships and meaning. These statements consist of a subject, a predicate, and an object (i.e., an article, has an author, Jane Smith). These statement parts are also referred to as a resource, a property, and a property value. Since there are three parts to RDF statements, they are referred to as triples. The predicate or property of an RDF statement defines the relationship between the subject and the object. RDF ontologies are sets of properties for a particular domain. For example, Darwin Core has an RDF ontology to express biological properties, 1 and EBUCore has an RDF ontology to express properties about audiovisual materials.2

Pulling apart the many issues involved in moving from XML to RDF is an exploration into the purpose of metadata, the tools available and their capabilities, and the various strategies that can be employed. Poupeau rightly states that XML provides structural logic in its hierarchical identification of elements and attributes, where RDF provides data logic declaring resources that relate to each other using properties.3 These properties are ideally all identified with single reference points (Uniform Resource Identifiers or URIs) rather than a description encased in an encoding. A source of honest confusion, however, is that RDF can be expressed as XML. Lassila's note regarding the Resource Description Framework specification from the World Wide Web Consortium (W3C) states, "RDF encourages the view of 'metadata being data' by using XML (eXtensible Markup Language) as its encoding syntax."4 So even though RDF can use XML to express resources that relate to each other via properties, identified with single reference points (URIs), RDF is itself not an XML schema. RDF has an XML language (sometimes called, confusingly, RDF, and from here forward called RDF/XML). Additionally, RDF Schema (RDFS) declares a schema or vocabulary as an extension of RDF/XML to express application-specific classes and properties.5 Simply speaking, RDF defines entities and their relationships using statements. There are various ways to make these statements, but the original way formulated by the W3C is using an XML language (RDF/XML) that can be extended by an additional XML schema (RDFS) to better define those relationships. Ideally, all parts of that relationship (the subject, predicate, object, or the resource, property, property value) are URIs pointing to an authority for that resource, that property, or that property value.

An additional concept worth covering is serialization. This term is used as a way to describe how RDF data is expressed using various formatting languages. RDF/XML, N-triples, Turtle, and JSONLD are all examples of RDF serializations.6 Describing something as being in RDF really means the framework of subject, predicate, object is being used. Describing something as being expressed in RDF/XML or JSON-LD means that the RDF statements have been serialized into either of those formatting languages. Using "RDF" to refer not only to the framework to describe something (RDF) but also the serialization of that description (RDF/XML) can easily muddle the discussion.

Other thoughts about the difference between XML and RDF or moving metadata from XML into RDF point to the difference in perspective and the change in thinking that is required to manage such a move. In an online discussion about RDF in relation to TEI (Text Encoding Initiative), Cummings talks about the need for both XML and RDF, using XML to encode text and RDF to extract that data and make it more useful.7 Yee, in her in-depth look at bibliographic data as part of the Semantic Web, points out that RDF is designed to encode knowledge, not information.8 The RDF Primer 1.0 also states "RDF directly represents only binary relationships."9 XML describes what something is by encoding it with descriptive elements and attributes. RDF, on the other hand, constructs statements about something using direct references—a reference to the thing itself, a reference to the descriptor, and a reference to the descriptor's value. As Farnel discussed in her 2015 Open Repositories presentation about the University of Alberta's move to RDF, they learned they were moving from a records-based framework in XML to a things-based framework in RDF.10 What is pointed out here time and again is something else Farnel discussed—moving from XML to RDF is not simply a conversion between encoding formats; it is a translation between two different ways of organizing knowledge. It involves understanding the meaning of the metadata encoded in XML and representing that meaning with appropriate RDF statements.

The tools most commonly employed for reworking XML into RDF are OpenRefine when accompanied by its RDF extension; a triplestore database such as OpenLink Virtuoso, 11 Apache Fuseki, 12 or Sesame13; Oxygen XML Editor14; and Protégé,15 an ontology editor. OpenRefine is, according to the website, "a powerful tool for working with messy data: cleaning it; transforming it from one format into another; and extending it with web services and external data."16 The RDF extension, called RDF Refine, allows for importing existing vocabularies and reconciling against SPARQL endpoints (web services that accept SPARQL queries and return results).17,18 SPARQL is similar to SQL as a language for querying a database, but the syntax is specifically designed to allow for querying data formatted in triple statements instead of tables with columns.19 Triplestore databases such as OpenLink Virtuoso can store and index RDF statements for searching as a SPARQL endpoint, offering a way to retrieve information and visualize connections across a collection of triples. Oxygen XML Editor has proven helpful in formulating eXtensible Stylesheet Language (XSL) transformations to move metadata from a particular XML schema or format into RDF/XML or other serializations such as JSON-LD (JavaScript Object Notation for Linking Data).20 Protégé is a tool developed by Stanford University that supports the OWL 2 Web Ontology Language and has helped to convert XML schemas to RDF ontologies and establish ways to express XML metadata in RDF. These tools provide the technical means to take metadata expressed in XML and physically reformat it to metadata expressed in an RDF serialization. What that reformatting also encompasses, however, is a review of the information expressed in XML and a set of decisions as to how to express that information as RDF statements.

Strategic approaches and ideas for handling data transformations into RDF have involved the XML schema or document type definition (DTD). These include Thuy, Lee, and Lee's approach to map an XML schema (the XSD) to RDF, associating simpleType's XSD in XML with properties in RDF, defining complexType's XSD in XML as classes in RDF, and handling a hierarchy of XML schema elements with top levels as domains and lower-level elements and attributes as container classes or subproperties in those domains.21 Thuy et al. earlier worked on a method to transform XML to RDF by translating the DTD using RDFS (ELEMENTs in the DTD are RDF classes or subclasses, ATTLISTs are RDF properties, and ENTITIES—preset variables in the DTD—are called up for use in RDF as encountered).22 Similarly, Hacherouf, Bahloul, and Cruz translate an XML schema into an OWL ontology.23 Klein et al. point out that while ontologies serve to describe a domain, XML schemas are meant to provide constraints on documents or structure for data so it can be advantageous to work out an RDF expression this way.24 Tim Berners-Lee puts it simply: "the same RDF tree results from many XML trees," meaning the same single statement in RDF (an article has an author Jane Smith) can be expressed in many ways in XML and can vary on the basis of the source of the XML, any schemas involved, and the people creating the metadata.[25] Transitioning from XML to RDF using the XML schema might serve to ensure all XML elements are replicated in RDF but does not necessarily establish the relationships meant by that XML encoding without additional evaluation. There is no single strategy that will always work to move XML metadata into RDF, even within the same set of tools (such as Fedora/Hydra) or the same area of concern (libraries, archives, or museums).

3. USE CASES FOR RDF

The following use cases explain approaches to transition to RDF taken from two differing perspectives. The first set describes efforts to express XML schemas or standards as RDF ontologies. The second set describes efforts by various library or cultural-heritage digital collections to transform metadata records into RDF statements. They also show that strategies to transform XML to RDF cannot occur without a shift in view from structure to relationships and, likewise, from descriptive encoding to direct meaning.

3.1 Moving an XML Schema/Standard to an RDF Ontology

As a graduate student at Kent State University, Mixter took on converting the descriptive metadata standard VRA Core 4.0 from an XML schema to an RDF ontology.26 Using the VRA Data Standards Committee Guidelines to ensure all minimum fields were included,27 Mixter mapped VRA XML elements and attributes to schema.org, FOAF, VoID, and DC Terms ontologies. This process is known as "cherry-picking," or combining various ontologies that already exist to represent properties or relationships (the predicates in RDF statements) as RDF instead of creating new proprietary RDF properties. Using OWL and RDFS as metavocabularies in Protégé, this created an ontology that could "retain the granularity required to describe library, archive, or museum items" of VRA Core 4.0's design in XML without being a straight conversion of VRA Core 4.0 from XML to RDF.28 The outcome was an XSLT stylesheet that was tested on VRA Core 4.0 XML records to produce that same information as RDF statements. One point that seemed to help in testing was the fact that all controlled vocabulary terms had reference identifiers in the XML (ready-made URIs). Something not discussed in the outcomes was that dates resulted in complex RDF (RDF statements that encompass additional RDF statements or blank nodes) and there was no discussion about this complexity or its effect on using those particular RDF statements.

VRA Core 4.0 now has an RDF ontology in draft form, with Mixter as one of its authors.29 The OWL ontology still points to schema.org, FOAF, and VoID for equivalent classes and properties, but everything is now named within a VRA RDF ontology and namespace and translates to such when VRA Core 4.0 XML is transformed to RDF.

Another case in the category of going from an XML standard to an RDF ontology is the development of the BIBFRAME model for bibliographic description from the Library of Congress. The BIBFRAME model is expressed as RDF. According to the BIBFRAME site, "in addition to being a replacement for MARC, BIBFRAME serves as a general model for expressing and connecting bibliographic data."30 MARC has its own format of expression with numbered fields and subfields but can be expressed or serialized in XML and is often shared that way. The BIBFRAME model, while revamping the way a bibliographic record is described on the basis of work, instance, authority, and annotation, also provides tools to transform records from MARC/XML to the RDF statements of BIBFRAME.31 A single namespace serves the BIBFRAME model and is explained as a long-term strategy to ensure namespace persistence over the next forty-plus years.32 The transformations produced from Library of Congress MARC records and local MARC records contain complex hierarchical RDF statements, particularly when ascribing authority sources to names, subjects, and types of identifiers.

As it is still a work in progress there are no tools making use of BIBFRAME records in RDF.

An additional example is the work happening with PBCore, the public broadcasting metadata standard managed by the Corporation for Public Broadcasting.33 Public broadcasting stations and other institutions across the United States provide descriptive, technical, and structural metadata for audiovisual materials using this XML standard. In Boston, WGBH's use of PBCore coincides with its digital asset management system, HydraDAM, built on Fedora 3 and the Hydra technology stack (based on Blacklight, Solr, and the Fedora Digital Repository).34 Fedora 3 does not natively support RDF statements as properties on objects like Fedora 4. Building off an interest to move HydraDAM to Fedora 4 and leverage RDF for metadata about audiovisual collections, WGBH began exploring transitioning the PBCore XML metadata standard into an RDF ontology. EBUCore, the European Broadcasting Union's metadata standard, is already expressed as an RDF ontology.35 A comparison between the XML standard of PBCore and the classes and properties expressed in EBUCore revealed that most PBCore elements were covered by the EBUCore ontology.36 Efforts are ongoing to offer PBCore 3.0 as an RDF ontology that uses EBUCore with the addition of a smaller set of properties along with a way to transform PBCore XML to PBCore 3.0 in RDF.37

The Hydra community, in an effort to help the transition from Fedora 3 with its XML binary files of descriptive metadata to Fedora 4 using RDF statements as properties on objects, is working on a recommendation and transformation to move descriptive metadata in MODS XML into RDF that is usable in Fedora 4.38 The MODS standard has a draft of an RDF ontology and a stylesheet transformation available,39 but the complex hierarchical RDF produced from this transformation is unmanageable with the current Fedora 4 architecture. The Hydra MODS and RDF Descriptive Metadata Subgroup is attempting to reflect the MODS elements in simple RDF statements that can be incorporated as properties on a Fedora 4 digital object.40 Led by Steven Anderson at the

Boston Public Library, this group is moving through MODS element by element, asking the question, "If you had to express this MODS element from your metadata in RDF today, how would you do that?" Participating institutions are reviewing their MODS records and exploring the possible RDF predicates that could be used to represent the meaning of that information. Some are even considering how to construct those RDF statements so that MODS XML can be re-created as close to the original MODS as possible (this is called "round tripping"). There are still questions as to whether every single MODS element will be reflected in this transformation, how exactly Fedora 4 will make use of these descriptive RDF statements, and if the original MODS XML will need to be preserved as part of the digital object in Fedora, but this group is recognizing that moving from Fedora 3 to Fedora 4 requires a major shift in thinking about descriptive metadata. This transformation tool is an effort to help make that transition possible.

The Avalon Media System is an open source system for managing and providing access to large collections of digital audio and video.41 It is built on Fedora 3 and the Hydra technology stack and uses MODS XML to store descriptive metadata. As development progresses and the available descriptive fields expand, maintaining the workflow to update XML records in Fedora and reindexing objects in the Hydra interface becomes increasingly complicated. Each time an update is made to descriptive information about an audiovisual item through the Avalon interface, the entire XML record for that object, stored as a binary text file, is rewritten in Fedora 3 and reindexed in Solr. In considering advantages to using Fedora 4, it appears that descriptive metadata properties stored in RDF are easier to manage programmatically (updating content, adding new fields, more focused reindexing) because descriptive information would not be stored in a single binary file but as individual properties on the object.

3.2 Turning XML metadata into RDF or Linked Data for publishing, search and discovery, and management

As Southwick describes the process, the library at the University of Nevada Las Vegas (UNLV) took a collection with descriptive records from CONTENTdm and published them as a single RDF Linked Open Data set.42 After cleaning up controlled vocabulary terms across collections and solidifying locally controlled vocabularies, they exported tab-delimited CSV records from CONTENTdm. These records were brought into OpenRefine with its RDF extension where they reviewed the data and mapped to various properties within the Europeana Data Model (EDM). Controlled vocabulary terms were in text form and had to be reconciled against a SPARQL endpoint, either locally from downloaded data or from the controlled vocabulary service, to gather the URIs to use as the object or value in the RDF statement. OpenRefine was then used to create RDF files that were uploaded to a triplestore (first Mulgara then OpenLink Virtuoso). This provided public access to the Linked Open Data set and a SPARQL endpoint for querying the data set. After publishing the data set they experimented with PivotViewer from OpenLink Virtuoso and RelFinder to see what kinds of connections and relationships could be visualized from the data as Linked Open Data. The outlined steps are clear and the outcomes are described, but interestingly the data set itself no longer appears to be available online.43 Although the UNLV use case relies on CSV instead of XML as the data source, the tools and workflows enlisted to transform the data set into RDF Linked Open Data are still applicable. OpenRefine can import XML just as it imports CSV, so this described case shows the tools that can be used and decisions to be made in processing that data into RDF statements.

In Oregon Digital,44 XML from Qualified Dublin Core, VRA Core, and MODS at two different institutions (University of Oregon and Oregon State University) were mapped as Linked Open Data and stored in a triplestore to be served up in a new web application using the Hydra technology stack.45 An inventory of metadata fields across all collections was first mapped to existing Linked Data terms, or properties (those with available URIs), then properties that were needed in the new web application but did not have available corresponding URIs were mapped to a newly devised local namespace for Oregon Digital. Any properties that were not used were kept in the original static XML file for the record as part of the digital object in Fedora. The focus here appears to be on mapping properties without as much detail provided on whether the objects were kept as text or mapped to URI values where possible. From the sample record provided the objects appear to be text and not URIs. The real power of this project is finding common properties to describe objects from diverse collections and institutions. What also comes out in the example mappings is the use of many different namespaces or ontologies (DC Terms, MARC Relators, but also MODS and MADS that produce complex RDF).

The University of Alberta also combined a variety of XML metadata from different sources into a new digital asset management system based on Fedora 4 and the Hydra technology stack, called the Education and Research Archive.46 Reporting on the experience at Open Repositories 2015, Farnel described the process as working in phases.47 Beginning with item types, languages, and licenses, then moving to place names and controlled subject terms, and finally person names and free-form subjects, they made multiple passes converting XML metadata into RDF statements and incorporating URIs whenever possible. They are combining all of this into a single data dictionary, 48 making use of several RDF ontologies to cover the various metadata properties that are being described about objects and collections.

University of California at San Diego (UCSD) has developed a local data model using a mix of external (MADS, VRA Core, Darwin Core, PREMIS) and local ontologies. They published a data dictionary and are working on a substantially different revision as part of the metadata workflow they use to bring digital objects into their digital asset management system from a variety of source metadata formats including XML.49 This allows metadata to be created from disparate source formats and makes it possible to bring them together as RDF for delivery, management, and preservation.

4. DISCUSSION

If metadata is in XML form and the desire is to express it as RDF, this is not merely a transformation from one XML schema to another. It is changing the expression of that data and changing its use. Having metadata in XML means information is encoded in a specific way that allows for interchange and sharing. Having metadata in RDF is making statements that have direct meaning and can be used independently.

There are different perspectives involved in metadata when approaching RDF: those that manage metadata standards (the XML standard side) and those that have metadata encoded using those XML standards (the data management side). Depending on the desired outcomes, the needs of these two perspectives can conflict. When managing a metadata standard the RDF transition tends to follow certain patterns:

- Transform an XML standard into a new RDF ontology
  - Examples: Dublin Core (DC), Darwin Core (DWC), MODS, VRA Core
- Establish a move to RDF that incorporates another existing ontology
  - Example: PBCore, Hydra community

From the data management side, the RDF transition means different patterns occur. These scenarios often start by reviewing the needed outcome, deciding how much metadata needs to be expressed in RDF, and what works best to get the metadata to that point. Cases include the following commonalities:

- Creating new search and discovery end user applications
    - Example: Oregon Digital, University of Alberta
- Publishing Linked Data sets
    - Example: UNLV, University of Alberta
- Managing metadata using software that supports RDF
    - Example: University of Alberta, UCSD, Hydra community

Conflicts are occurring when the needed outcome on the data management side is not supported by the RDF ontology transitions that have occurred for the XML standards being used. An example of this is how RDF is handled in Fedora 4. When RDF is complex (the object of one statement is another entire RDF statement), Fedora produces blank nodes as new objects within the repository. While not technically problematic, descriptive metadata with complex RDF can result in a situation where a digital object ends up referencing a blank node that then points to, for example, a subject or a genre. This subject or genre has been created as its own object within the digital repository even though that subject or genre is only meant to provide meaning for the digital object. MODS RDF produces this complexity and thus is not workable to use with Fedora 4. In contrast, other standards such as DC or DWC in RDF produce simple statements that Fedora 4 can apply to a digital object without any additional processing.

Complications in transitioning from XML to RDF also occur when the original XML does not include URIs or authority-controlled sources. Converting this metadata to RDF can mean locally minting URIs or bringing data over as literals (strings of text) without using URIs at all. Ideally, the result is somewhere in the middle with externally controlled vocabularies incorporated as much as possible and literals or locally minted URIs only used where absolutely necessary. Translating strings to authoritative sources is intensive work. If the XML standard cannot be expressed as a single RDF ontology, work is further complicated by the need to map XML elements to different RDF ontologies using logic that is often decided locally.

While it is possible to transition XML to RDF, the process is not uniform and the pathway involves a lot of labor. Potential alleviators for this labor might involve a more user-centered approach by XML standard bodies to consider the ways their standards can be used when translated into RDF ("users" in this context meaning the users of the standards, not the end users searching and discovering digital content). Triplestores can manage queries for complex RDF, but digital repository systems are not there yet. Those that support RDF for description of objects do so on the basis of simple property statements. A complex RDF ontology is going to be a challenge to support over time.

Another way to progress forward is for the data management side of the equation to focus efforts on showing, in an end user search and discovery format, what is currently possible when XML is transitioned into RDF. Published Linked Data sets need to have interfaces for access and use, showing the value of what is currently available and any needs or gaps that remain. Libraries and cultural-heritage organization engaged in this work should also openly share the processes that both work and do not work so others contemplating this transformation can consider how to forge ahead themselves. Libraries and cultural-heritage organizations moving metadata from XML to RDF should provide feedback to XML standard bodies regarding the usefulness or complications of any RDF transitional help an XML standard might provide.

Technologies for incorporating RDF into web applications and truly connecting triples across the web also require further work. Triplestores have so far been the main way to expose data sets but have not been incorporated into common library or cultural-heritage end user search and discovery web applications. Additionally, triplestore use does not seem to extend to management or long-term storage of complete data about digital objects. There seems to be a decision to either reduce the data stored in a triplestore down to simple statements or use the triplestore more like an isolated index or SPARQL endpoint only and manage the complete metadata record separately (in a static file containing text or in a separate database). That aligns triples in RDF more with relational database storage than with catalog records. Triple statements focus on relationships and not the complete unique details of the thing being described. Triplestores can handle complex hierarchical RDF graphs and provide responses on the basis of queries against those complexities, 50 but triplestores do not appear to be taking over as either the main search and discovery mechanism for online digital resources or for digital object management. Software using RDF natively is also not currently widespread. A project such as the BIBFRAME Initiative that plans to incorporate RDF needs to make sure the complexity of its data model in RDF is manageable by any tools it produces and that it is possible for vendors and suppliers to encompass the data model in their software development.

5. CONCLUSION

The reasons for deciding metadata should transition to RDF are just as important as determining the best process for implementing that transition. Reasons for transitioning to RDF are conceptually based around making data more easily shareable and setting up data to have meaning and relationships as opposed to local static description that requires programmatic interpretation. The use cases outlined in this article show the reality does not quite yet match the concept. Transitioning an XML standard to RDF does not make that data more shareable or more easily understood unless there are end user applications for using that data in RDF. Publishing Linked Data involves going through transitional steps, but the endpoint seems to be more of a byproduct. The real goal is going through the process of producing Linked Data to learn how that works. Self-contained projects that aim to express collections in RDF for the purpose of a new search and discovery interface are more successful in implementing RDF that has that new level of meaning and relationship. Beyond the borders of these projects, however, the data is not being shared or used.

The use cases described above show some examples of what is happening now when transitioning from XML to RDF. Approaches include XML standards converting to RDF expression as well as digital collections with metadata in XML that have an interest in producing that metadata as RDF. Software that incorporates RDF is still developing and maturing. Helping that process along by providing a pathway from XML to functionally usable RDF improves the chances of the Semantic Web becoming a real and useful thing. It is vital to understand that transitioning from XML to RDF requires a shift in perspective from replicating structures in XML to defining meaningful relationships in RDF. Metadata work is never easy, and for metadata to move from encoded strings of text to statements with semantic relationships requires coordination and communication. How best to achieve this coordination and communication is a topic worth engaging as the move to use RDF, produce Linked Data, and approach the Semantic Web continues.

BIBLIOGRAPHY

1. Berners-Lee, Tim. "Linked Data." Linked Data - Design Issues, June 18, 2009.

2. Estlund, Karen, and Tom Johnson. "Link It or Don't Use It: Transitioning Metadata to Linked Data in Hydra," July 2013.

3. Farnel, Sharon. "Metadata at a Crossroads: Shifting 'from Strings to Things' for Hydra North." Slideshow presented at the Open Repositories, Indianapolis, Indiana, 2015.

4. Hacherouf, Mokhtaria, Safia Nait Bahloul, and Christophe Cruz. "Transforming XML Documents to OWL Ontologies: A Survey." Journal of Information Science 41, no. 2 (April 1, 2015): 242–59.

5. Klein, Michel, Dieter Fensel, Frank van Harmelen, and Ian Horrocks. "The Relation between Ontologies and XML Schemas." In Linköping Electronic Articles in Computer and Information Science, 2001.

6. Lassila, Ora. "Introduction to RDF Metadata." W3C, November 13, 1997.

7. Manola, Frank, and Eric Miller. "RDF Primer 1.0, Section 2.3 Structured Property Values and Blank Nodes." W3C Recommendation, February 10, 2004.

8. Mixter, Jeff. "Using a Common Model: Mapping VRA Core 4.0 Into an RDF Ontology." Journal of Library Metadata 14, no. 1 (January 2014): 1–23.

9. Poupeau, Gautier. "XML vs RDF: logique structurelle contre logique des données (XML vs RDF: structural logic against logic data)." Les Petites Cases, August 29, 2010.

10. "RDF and TEI XML," October 13, 2010.

11. Southwick, Silvia B. "A Guide for Transforming Digital Collections Metadata into Linked Data Using Open Source Technologies." Journal of Library Metadata 15, no. 1 (March 2015): 1–35.

12. Thuy, Pham Thi Thu, Young-Koo Lee, and Sungyoung Lee. "A Semantic Approach for Transforming XML Data into RDF Ontology." Wireless Personal Communications 73, no. 4 (2013): 1387–1402.

13. Thuy, Pham Thi Thu, Young-Koo Lee, Sungyoung Lee, and Byeong-Soo Jeong. "Transforming Valid XML Documents into RDF via RDF Schema." In Next Generation Web Services Practices, International Conference on, 0:35–40. Los Alamitos, CA: IEEE Computer Society, 2007.

14. "XML RDF." W3Schools. Accessed September 30, 2015.

15. Yee, Martha M. "Can Bibliographic Data Be Put Directly onto the Semantic Web?" Information Technology and Libraries 28, no. 2 (March 1, 2013): 55–80.

Chapter 9

Applications of Web 2.0 tools in IIT Libraries in India: A Study

Dr. Shalini R. Lihitkar[1], Vaibhav P. Manohar[2]

[1]Deptt. of Library and Inf. Science Rastrasant Tukadoji Maharaj Nagpur University Nagpur (India)

[2]B. D. College of Engineering, Sewagram, Distt. Wardha

ABSTRACT

Web 2.0 is a second generation website that deals with the ability, of people being able to exchange information online. Web 2.0 based technologies have provided more opportunities and opened up new ways of communicating and collaborating. The transition to Web 2.0 is based on 'interactive' use of web. Users can carry on interaction with others. People can collaborate on the same document at fairly much at the same time. Users can post thoughts and creative expressions, and there are places for other to leave comments. Web 2.0 has created a platform for increasing interpersonal content facilitation, sharing, creativity, service delivery, innovation, and collaboration. Organizations have been taking advantage of Web 2.0 technologies in many ways- by providing a better communication channel for all stakeholders for sharing their experiences & collaborative working and introducing innovative way of marketing products and services etc. Web 2.0 technologies have played a significant role in our ability to keep up with the changing needs of library users

This article focuses on the impact of web 2.0 on libraries and information services of IIT Libraries in India. Also discussed the Web 2.0 tools used by IIT libraries for providing library 2.0 services to their users.

KEYWORDS

Web 2.0, IIT, Libraries, Tools, India, Blog, Twitter, Youtube, RSS Feed social Networking, Mash-up, Cloud computing, Instant messaging, chatting, Flickrs

1. INTRODUCTION

The quality of education system in any society is a measure of its development and competitiveness. In the present age of knowledge explosion, where new technologies are being developed at a rapid pace, their assimilation with the existing knowledge, development of infrastructure, imparting of education and training to cater to the growing requirement for the qualified and trained manpower for the operationalization of technologies are the responsibility of the Information Technology institutes, Management institutes, and Science institutes education system.

The top institutions of India like Indian Institute of Technology (IIT) libraries plays a significant role in the Engineering education system. For a student seeking a graduate degree, it is largely an information sources center that supplements class room and outside learning. The IIT libraries serves as a sources of information for new technological developments and their application. Thus such libraries are expected to identify, acquire, organize, store, retrieve and disseminate information not only on the specific subject covered by the curriculum, but also on the current directions of technological advancement.

2. RESEARCH PROBLEMS

The development of emerging technologies has resulted in a rapid increase of information and communication technology (ICT) tools. Web 2.0 technologies have played a significant role in our ability to keep up with the changing needs of library users. Web 2.0 based technologies have provided more opportunities and opened up new ways of communicating and collaborating. So, It is very essential to know the current status of Web 2.0 services provided by IIT Libraries in India. The library professionals need to have knowledge of such web 2.0 tools so that they can provide prompt reference and other services to the users.

The study of "Applications of Web 2.0 in IIT Libraries in India" can encourage the librarians to apply the Web 2.0 tools effectively in their libraries. As well as this study encourage the users to utilize such services which is very useful for their higher study. The present study will prove to be a guide to know various aspects of Web 2.0 tools for the users as well as researchers and LIS professionals.

3. REVIEW OF LITERATURE

Review of literature is an early step for conducting research. It enables to avoid the duplication of research work and broadens the understanding of the research problem. This is an overview of the literature which provides a background to the area of research under consideration to various research studies related to the present research problem "Applications of Web 2.0 in IIT's, Libraries in India: A

Study". The relevant data was collected from various sources like books, journals, magazine, proceeding, thesis and the web articles in open access and peer reviewed journals. The emergence of the new Web 2.0 applications and the research conducted on them are discussed, and a brief introduction to the application used in this study is included.

Bhatt, Chandra, and Denick5 showed that how can Web 2.0 tools function as information awareness tools for faculty and students in an academic community? Why is 'information awareness' so crucial? He wished that more faculty and students realize the potential of Web 2.0 applications in keeping up- to – date with new information and information resources. He mentioned that how various Web 2.0 applications like Blogs and RSS feed, Social bookmarking, friend feed, Facebook are being used at Drexel University. He concluded that the Web 2.0 technology are having an important impact on library services. Morange, Joel (2010)34 discussed about Web 2.0 and controversial point about the same as well as with virtual communities. Dasgupta (2012)6 discussed the various groups of social network and their behavior. And also discussed the benefit of the social networking. Lihitkar, (2010)27 discussed the importance of the web 2.0 tools in the present libraries. She concluded that Weblog is becoming increasingly popular in higher education. The role of an instructor is very important in making it an effective tool. There are many ways in which Blogs and RSS feeds can be used in academic environment to promote and market services and resources to faculty and the students. Librarians are using Blogs in unique and diverse ways for e.g. in personal management, class announcements, posting student work for viewing etc. Lihitkar (2011), described in the research paper that the concept of virtual reference services and Instant messaging, with the salient features of instant messaging services and their usefulness as a reference services in the libraries. As well as overviewed various IM softwares available for the use.

4. OBJECTIVES OF THE STUDY

The basic aim of the study was to find out the applications of Web 2.0 in Indian Institute of Technology (IIT) Libraries in India. The present study were carried out with the following objectives:

- To know the concept of Web 2.0 and its applications in providing library services to users.
- To make a survey in order to know the applications of Web 2.0 among the IIT Libraries in India.
- To examine the characteristic features of Web 2.0 tools used in IIT Libraries in India.
- To offer suggestions regarding effective use of Web 2.0 Technology in Libraries.

5. SCOPE AND LIMITATIONS

The scope of the study was limited to IIT's Libraries in India. The Indian Institutes of Technology (IITs) are a group of autonomous engineering institutes of higher education, which has declared as "Institutions of National Importance" by the Govt. of India. The IITs are governed by HRD, Govt. of India under the Institutes of Technology Act, 1961. Indian Institute of Technology are centrally funded institutes which are looked after by the Government of India. At Present, there are only 16 IIT's college libraries all over India in which seven IIT colleges are old and 08 IIT colleges are started in the year 2008–09 and 01 IIT (BHU) Varanasi granted recognition as IIT as a national important and started in the year 2011.

Table No. 1. List of IIT's in India (www.wikipedia.com)

Sr. No .	Name	Short Name	Est. year	Recogniti on as IIT	City/Town	State/UT	Questionni ars Distributed	Respon se Receive d
1	IIT Kharagpur	IITKG P	195 0	1951	Kharagpur	West Bengal	√	√
2	IIT Bombay	IITB	195 8	1958	Mumbai	Maharasht ra	√	√
3	IIT Madras	IITM	195 9	1959	Chennai	Tamil Nadu	√	√
4	IIT Kanpur	IITK	195 9	1959	Kanpur	Uttar Pradesh	√	√
5	IIT Delhi	IITD	196 1	1963	New Delhi	Delhi	√	√
6	IIT Guwahati	IITG	199 4	1994	Guwahati	Assam	√	√
7	IIT Roorkee	IITR	184 7	2001	Roorkee	Uttarakha nd	√	√
8	IIT Ropar	IITRP R	200 8	2008	Rupnagar	Panjab	√	√
9	IIT Rajasthan	IITJ	200 8	2008	Jodhpur	Rajasthan	√	√
10	IIT Bhubanesw ar	IITBB S	200 8	2008	Bhubanesw ar	Orissa	√	√
11	IIT Hyderaba	IITH	200 8	2008	Hyderabad	Andhra Pradesh	√	√
12	IIT Gandhinag ar	IITGn	200 8	2008	Gandhinag ar	Gujarat	√	√
13	IIT Patna	IITP	200 8	2008	Patna	Bihar	√	√
14	IIT Indore	IITI	200 9	2009	Indore	Madhya Pradesh	√	√
15	IIT Mandi	IIT Mandi	200 9	2009	Mandi	Himachal Pradesh	√	√
16	IIT –BHU	IIT-BHU	201 2	2012	Varanasi	IIT -BHU	√	x
							16	**15**

6. RESEARCH METHODOLOGY

The descriptive method of research includes questionnaire and interview tool has been followed for current research study. The questionnaire was designed with questions on staff, infrastructure, resources available and Web 2.0 services provided by the IIT Libraries in

India. These questionnaires were sent to all IIT Libraries in India. Required data collected from 16 IIT's Libraries in India. Out of 16 Libraries 15 respondents were received. The response rate is 93.75% , in which some data are personally collected, some response received by the post and some respondents sent questionnaires through Email and data also collected from the respective website.

7. WEB 2.0 TECHNOLOGY

Web 2.0 is not a technological change or a change in programming language. Web 2.0 refers to a revolution in how the internet is used. Web 2.0 applications concentrate on sharing content with the masses to let others comment on it or to create new content of their own. In this technology anyone and everyone can publish their work on the internet with ease. Web 2.0 is a socially updated version of Web 1.0. While Web 1.0 was limited to just reading. A Web 2.0 site gives its users the free choice to interact or collaborate with each other in a social media dialogue as creators of user-generated content in a virtual community.

Web 2.0 technology started in the year 2002 to 2009 in which the various companies included like Facebook, LinkedIn, Group on etc. Web 2.0, in particular, has brought prominent changes that affect internet users and application developers. The main agenda of Web 2.0 is to connect people in numerous new ways and utilize their collective strengths.

The term "Web 2.0" was coined in 1999 by Darci Di Nucci who was a consultant on electronic information design. This term did not come out until 2003. In 2004, the term began to rise in popularity O'Reilly Media and Media Live hosted the first Web 2.0 video conference. The term Web 2.0 closely associated with Tim O'Reilly because of the O'Reilly media Web 2.0 conference in 2004. Tim O'Reilly, one of the first to use the term Web 2.0, describes it as viewing the web as a platform to distribute content. Web 2.0 allows user to do much more than just retrieve information. They provide the user with more user-interface, software and storage facilities, all through their browser. User can provide the data that is on a Web 2.0 website and exercise some control over the data.

"Web 2.0" refers to web development and web design that facilitates interactive information sharing, interoperability, user-centered design and collaboration on the www. Examples of web 2.0 include web-based communities, hosted services, web applications, social networking sites, video-sharing sites, wikis, blogs, mash-ups and YouTube, Social bookmarking, tagging, folksonomies etc. Web 2.0 technologies a second generation of services available on the web that allows its users to interact with other users or to change website content. Web 2.0 is not a technological change or a change in programming language. Web 2.0 refers to a revolution in how the internet is used. Web 2.0 applications concentrate on sharing content with the masses to let others comment on it or to create new content of their own.

John Robb wrote: "Web 2.0 is a system that breaks with the old model of centralized Web sites and moves the power of the Web/Internet to the desktop." Web 2.0 is defined as "the second generation of the www in which content is user generated and dynamic,"

According to Tim O'Reilly "Web 2.0 is the business revolution in the computer industry caused by the move to the Internet as a platform, and an attempt to understand the rules for success on that new platform" 12 Tim Berners-Lee (2006) defined that "Web 2.0 is a piece of jargon"

Linda Young (2008) defined Web 2.0 as "the second generation of the Web, which enables people with no specialized technical knowledge to create their own websites, to self-publish, create and upload audio and video files, share photos and information and complete a variety of other tasks."

8. DIFFERENCE BETWEEN WEB 1.0 AND WEB 2.0 TECHNOLOGY

Table No.2. Difference between Web 1.0 and Web 2.0

Sr. No.	Web 1.0	Web 2.0
1	**Tim Berners-Lee**	**Darcy DiNucci** (a Consultant on electronics information design)
2	Read Only	Read and Write
3	Static web pages	Dynamic web pages
4	Content published by the webmaster	Open content to be used and reuse, wiki
5	No user participation,	Active user participation
6	Not often updated	(usually) Frequently updated
7	Communication via Email	Communication via Facebook, Instant messaging, Blog, Twitter, chatting
8	Software on PC	Software on PC, Tablet, mobile, laptab
9	Wire	Wireless
10	Directories (Taxonomy)	Tagging (folksonomy)
11	Content Management System	Wiki
12	Britannica online	Wikipedia
13	Personal website	blogging

9. CHARACTERISTIC OF THE WEB 2.0 TECHNOLOGY

Web 2.0 websites allow users to do more than just retrieve information. The Web 2.0 offers all users the same freedom to contribute. According to Best, the characteristics of Web 2.0 are:

- Active Users participation
- Open content to be used and reused
- Decentralization
- Modularity
- User Control

- Rich user Interactions.
- Dynamic web pages
- Openness
- Freedom
- Collective Intelligence

10. ADVANTAGES OF WEB 2.0 TECHNOLOGY [12]

There are following advantages of the Web 2.0 technology.

- To support innovation
- Creativity
- Collaboration
- Information Sharing
- Substantial Cost and time saving on projects.
- Confidentiality and privacy of data
- Flexibility to participate from anywhere
- Ability to conduct multilingual studies and automatic generation of report, table, chart and transcript

11. TYPES OF WEB 2.0 TOOLS:

Web 2.0 tools are internet based services which includes social networking sites, Blogs, Twitter, Wikis, Instant Messaging, Social Networking, Social Bookmarking, Skype, Facebook, Mash-ups, Cloud computing, Tagging, Folksonomy, MySpace etc. These services aim to accelerate creativity, support information sharing, online collaboration and networking among the users. The use of Web 2.0 tools has grown considerably in the education sector in the last few years.

All these tools are Social, Participatory, Collaborative, Easy to use, Create community environment- because they are online communities. There are so many tools as mention in the following Figure.

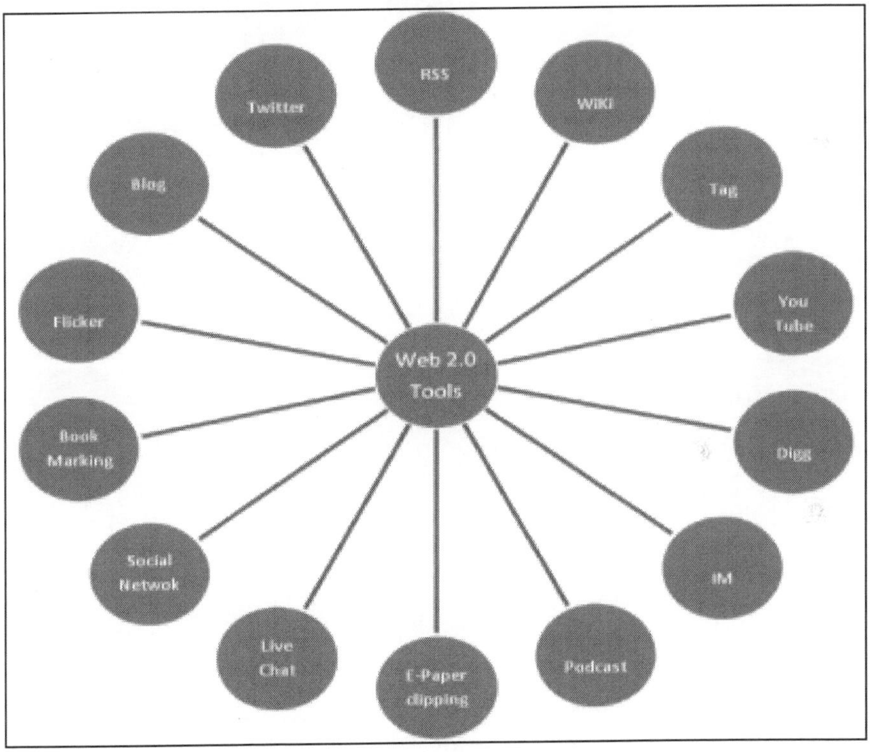

Figure 1. Web 2.0 Tools

There are total 16 IIT's Libraries in India and response received was 15 libraries. But out of 15 libraries, we found that only 12 libraries is using Web 2.0 technology and providing library 2.0 services to users. Following table shows the details.

 11. Types of Web 2.0 tools:

Table No. 3. Use of Web 2.0 tools In the IIT's Libraries in India

Sr. No	Name of the IIT'S, Libraries in India	Social Networking	Blog	Twitter	YouTube Video	RSS Feed	Instant Messaging	Social Bookmarking	Tagging	Delicious	Flickr	Podcasting	Folksonomy	Mash-up	Cloud Computing	Wiki	Virtual tour	Total Web 2.0
1	IIT, Delhi	√	x	√	x	√	√	x	√	x	x	x	x	x	x	√	x	**6**
2	IIT, Kharagpur	√	√	x	√	√	√	x	x	x	x	x	x	x	x	x	√	**7**
3	IIT, Madras	√	√	√	x	√	√	x	x	x	√	x	x	x	x	√	x	**7**
4	IIT, Roorkee	√	√	√	√	x	x	x	√	x	√	x	x	x	x	x	x	**6**
5	IIT, Kanpur	√	x	x	√	√	x	x	√	x	√	x	x	x	x	x	x	**5**
6	IIT, Bombay	√	√	√	√	√	√	√	√	√	√	√	x	x	x	√	x	**12**
7	IIT, Indore	√	√	√	x	√	x	x	x	x	x	x	x	x	x	√	x	**5**
8	IIT, Jodhpur	√	√	x	x	x	x	x	x	x	x	x	x	x	x	x	x	**2**
9	IIT, Ropar	√	x	x	x	√	x	x	x	x	x	x	x	x	x	x	x	**2**
10	IIT, Mandi	√	x	x	√	x	√	x	√	x	x	x	x	x	x	x	x	**5**
11	IIT, Hyderabad	√	√	√	x	√	√	√	√	x	x	x	x	x	√	√	x	**10**
12	IIT, Gandhinagar	√	x	√	x	√	√	x	x	x	x	√	x	x	x	x	x	**6**
	Total	**12**	7	**7**	**5**	**9**	**7**	**2**	**6**	**1**	**4**	**2**	**0**	**0**	**1**	**5**	**1**	

Table 4: shows that Social Networking site have been used by maximum 12 (100%) IIT's, libraries in India. Social Networking tool is the most popular tool to communicate with users and other staff members. Social Networking tools includes the Facebook, LinkedIn, MySpace, Gmail, Word Press, Blogs, Twitter, IM etc. The use of these tools is increasing among the students nowadays. Social networking got first rank.

RSS (Really Simple Syndication) tool was used by maximum 09 (75%) IIT's, libraries in India. The popularity of RSS may be attributed to its simplicity and easy to use feature. The users can subscribe to the RSS feeds which cater to their academic and research needs. For example, the library of IIT Bombay provides a variety of RSS news feeds to keep users up to date with what is going on around the IIT Bombay. The news feeds are available from department of library news. RSS feed got second rank in using IIT Libraries.

Table No. 4. Analysis of use of Web 2.0 tools In the IIT's Libraries in India

Sr. No	Web 2.0 tools	Web 2.0 tools used in IIT's Libraries in India Out of 12	Persentage %	Ranking
1	Social Networking	12	100 %	1
2	RSS Feed	09	75%	2
3	Blog	07	58.33 %	3
4	Twitter	07	58.33%	3
5	Instant Messaging	07	58.33%	3
6	Tagging	06	50%	4
7	YouTube, Video sharing	05	41.66%	5
8	Wiki	05	41.66%	5
9	Flickr	04	33.33%	6
10	Social Bookmarking	02	16.66%	7
11	Podcasting	02	16.66%	7
12	Cloud Computing	01	8.33%	8
13	Delicious	01	8.33%	8
14	Virtual tour	01	8.33%	8
15	Folksonomy	00	Nil	
16	Mash-up	00	Nil	

The Blogs, Twitter and Instant Messaging were used by maximum 07 (58.33%) IIT's, libraries in India. The use of these tool is being getting immense popularity. These are very easy to maintain and do not require much of time or effort. Another strong feature is availability of free hosting software. These tools got 3rd rank in using IIT Libraries.

The Tagging was adopted by 06 (50%) IIT's, libraries in India. Tagging tools got 4rd rank in using IIT Libraries. This tool is very important for the researcher. While searching the article of our interest, we can tag the article for further use. Tag is like a bookmarking tool. Everyone can download the article and tag if favorite.

The YouTube and wiki tools was applied by minimum 05 (41.66%) IIT's, libraries in India. These tools got 5rd rank in using IIT Libraries.The Flickr was adopted by 04 (33.33%) IIT's, libraries in India. Flickr got 6rd rank in using IIT Libraries.

Podcast is a digital media file or a related collection of such files which is distributed over internet using syndication feeds for playback on a portable media player and personal computers. The Podcast / Vodcast was the least used tool it was found that just only 02 (16.66%) IIT's, libraries in India used it. These tools offer users a chance to listen or view to the files whenever or wherever they want to. Social Bookmarking is also very famous web 2.0 tool. We can bookmark any article or any document and share among the interested friends.

The Podcast and social bookmarking were adopted just only 02 (16.66%) IIT's, libraries in India that means IIT Bombay and IIT Gandhinagar.

The Delicious, Cloud computing, mash-up were the least used tools- it was observed that each tools used by only 01 IIT library in India. None of the Libraries are using folksonomy and mash-up tools.

It is observed that Social Networking were more used 12(100%) whereas cloud computing was less used as well as Mash-up and folksonomy were not used in the IIT's, libraries in India.

Table No. 5. Analysis of Way of knowledge gained by LIS professionals about use of Web 2.0 tools in IIT's, Libraries in India.

S N	Way of knowledge for LIS professionals about web 2.0	IIT Libraries in India
		IIT
1	Training provided by the organization	0
2	Fund provided by the organization for training	0
3	Self-Motivated	11
4	Friends encourage	01

As indicated in Table 5, it is noted that Maximum 11(91%) IIT's, librarians in India were received the knowledge about web 2.0 by self-motivation. The librarians themself done practice of web 2.0 tools and try to implement these tools for providing the Library 2.0 services.

12.PURPOSE OF WEB 2.0 TOOLS USED IN IIT LIBRARIES IN INDIA

Table No.7. Indicates that 08 (66.66%) of IIT's, Librarians in India were in opinion that web 2.0 tools are useful for having better communication with the users. 07 (58.33%) Librarians given their opinion in favor of web 2.0 tools are important to provide quick services to the users. Another 07 (58.33%) Librarians said that web 2.0 tools are essential for developing digital library services, And 5 (41.66%) Librarians of IIT's, Libraries in India given their views in favour of web 2.0 tools are more useful to update the knowledge of the users.

Table No. 6. Purpose of Web 2.0 Tools used in IIT Libraries in India

Sr. No.	Web 2.0 tools & Purpose used in IIT Libraries	IIT Delhi	IIT	IIT Madras	IIT Roorkee	IIT Kanpur	IIT Bombay	IIT Indore	IIT	IIT Ropar	IIT Mandi	IIT	IIT Gandhi	
1	**Social Networking** 1.Keep in touch with family & Friend 2. To share information using a variety of features like pictures, Video clip 3. To Create profile in details 4. To suggest information needs on the Librarian wall on Facebook 5. To Keep in touch CAS 6. To promote Library Services	- - - √ √ -	- - √ - √ -	- - - - √ -	- - - √ √ -	- - √ √ √ -	- - - - √ -	- - - - -	√ - - - - -	√ - - √ √ -	√ - - - √ -	- - - √ √ -	- - - - √ √	**03** **0** **02** **05** **10** **01**
2	**Blog** 1. For library related News & Events 2. For new releases books, Databases, video etc 3.Marketing of Lib. & Inf. services 4. As a discussion Forum 5. Collection Development 6.Appeal to special interested group	- - - - - -	√ - - - - -	√ - - - - -	√ - - - - -	- - - - - -	- - - - √ -	√ - - - - -	- - - - √ -	√ - - - - -	- - - - - -	- - - - √ -	- - - - - -	**05** **0** **0** **0** **03** **0**
3	**Twitter** 1. For library related News & Events 2. For discussion 3. To stay connected with other librarians, Friends & Students 4. To interact with vendor 5. For reference services	- - √ - √	- - - - -	√ - - - -	√ - √ - -	- - - - -	√ - √ - -	- - - √ -	- - - - -	- - - - -	- - - - -	- - √ - √	√ - - - -	**04** **0** **04** **01** **02**
4	**Tagging** 1.To enter the specific location 2. To Edit data 3. To collect data	√ - -	- - -	- - -	√ √ √	√ - -	√ √ √	- - -	- - -	- - -	√ √ √	√ √ √	- - -	**06** **04** **04**

5	**Social Bookmarking**													
	1. To use keyword shortcut for some of favorite links	-	-	-	-	-	-	-	-	-	-	-	-	.0
	2. To add personal Collections	-	-	-	-	√	-	-	-	-	-	√	-	02
	3. To save the time and efforts in searching	-	-	-	-	√	-	-	-	-	-	-	-	01
	4. To maintain own webpage link	-	-	-	-	√	-	-	-	-	-	√	-	02
	5. To search specific topic	-	-	-	-	-	-	-	-	-	-	-	-	0
6	**Youtube**													
	1. For Entertainment	-	-	-	-	-	-	-	-	-	-	-	-	0
	2. To show the library services	-	√	-	√	√	√	-	-	-	√	-	-	05
	3. To Find Library resource services	-	√	-	√	-	√	-	-	-	-	-	-	02
	4. To discover new ways to implement technology in the library	-	-	-	-	-	-	-	-	-	-	-	-	0
	5. To teach library etiquettes	-	-	-	-	-	-	-	-	-	-	-	-	0
7	**RSS**													
	1. For library related News & Events	√	√	-	-	√	√	√	-	√	-	√	√	08
	2.To keep students update on key information / Research article	-	-	-	-	-	-	-	-	-	-	-	-	0
	3. For e-journals / Online Databases	-	√	√	-	-	√	-	-	-	-	√	-	04
	4. To save the time of researcher/ Students and faculty	√	-	-	-	√	-	√	√	-	-	√	√	06
	5. Organizational use	√	-	√	√	√	√	-	-	√	-	√	√	08

8	**Wiki**													
	1. For to Searching Information	√	-	√	-	-	√	-	-	-	-	√	-	**04**
	2. For to allow users to create, edit and link web pages easily.	√	-	√	-	-	√	-	-	-	-	√	-	**04**
	3. Use as Research tool	-	-	-	-	-	-	-	-	-	-	-	-	**0**
	4. To enable user to upload the files	-	-	-	-	-	-	-	-	-	-	-	-	**0**
	5. Knowledge Sharing	√		√	-	-	√	√	-	-	-	√	-	**05**
9	**Instant Messaging**													
	1. Instant clarification for the Questions	√	√	√	-	-	-	-	√	-	-	√	√	**06**
	2. For Library offering text based chat	√	√	√	-	-	√	-	√	-	√	√	√	**08**
	3. For online meeting	√	√	-	-	√	-	-	-	-	-	√	√	**05**
	4. For providing Virtual Reference Service	√	-	√	-	-	-	-	√	-	√	√	-	**05**
	5. Guidance about resources	√	√	√	-	-	-	-	-	-	-	-	-	**03**
10	**Podcasting**													
	1. For current News about what is going on in the library	-	-	-	-	-	√	-	-	-	-	-	√	**02**
	2. For highlighting about new resources	-	-	-	-	-	-	-	-	-	-	-	-	**0**
	3. To share information to anyone	-	-	-	-	-	-	-	-	-	-	-	-	**0**
	4. Speeches of important personality	-	-	-	-	-	-	-	-	-	-	-	-	**0**
	5. Library orientation tour	-	-	-	-	-	-	-	-	-	-	-	-	**0**

Table No. 7, shows that maximum 12 (100%) IIT's, Library professionals in India were co-operating to the users for providing web 2.0 services.

Table No.8, highlights that maximum 09 (75%) Librarians of IIT's Libraries in India were facing technical problems while using Web 2.0 tools, and 03 (25%) of them are facing the problem that means not receiving the proper training about how to use web 2.0 tools.

The table No 9 shows the Ranking of 12 IIT's, Libraries in India based on checkpoints used. IIT Library Bombay got 1st Rank, IIT library, Hyderabad got 02 rank application Index. IIT Library, Kharagpur, Madras, got 3rd in Rank, Roorkee, Delhi and Gandhinagar got 4 th rank.Kanpur, Indore , mandi got 5 rank and Jodhpar and Ropar got 6 rank,.

17. SUGGESTIONS

The surveyed the subsequent analysis of the data and the findings of the study have enabled the researcher to provide some practical suggestions for improving the use of Web 2.0 tools for library services expected from IIT's, libraries in India as follows:

Table 7. Views of IIT Librarians and Attitudes of the library professionals for using Web 2.0 tools in Libraries in India

Sr. No	Name of the IIT's , Libraries in India	Views of Librarians for using Web 2.0 tools				Attitudes of IIT Library Professional	
		For Quick Services to the	For Digital Library	For Update the Knowledge of the Users	For better Communication with the Users	LIP co-operating for providing services to the	LIP Not co-operating for providing services to the
1	IIT Library, Delhi	√	√	√	√	√	x
2	IIT Library, Kharagpur	√	x	x	√	√	x
3	IIT Library, Madras	x	√	x	√	√	x
4	IIT Library, Roorkee	√	x	√	√	√	x
5	IIT Library, Kanpur	x	x	√	x	√	x
6	IIT Library, Bombay	√	√	x	x	√	x
7	IIT Library, Indore	√	√	x	√	√	x
8	IIT Library, Jodhpur	x	√	x	√	√	x
9	IIT Library, Ropar	x	x	x	√	√	x
10	IIT Library, Mandi	√	√	x	√	√	x
11	IIT Library, Hyderabad	√	x	√	x	√	x
12	IIT Library, Gandhinagar	x	√	√	x	√	x
	Total	7	7	5	8	12	0

Table No. 8. IIT's, Librarian in India faced the problems while using Web 2.0 tools.

Sr. No	Name of the IIT's, Libraries in India	Problems faced				
		I do not know how to use the tools	Not received proper training	Technical problems	Lack of access to the tools	Lack of support of L & P
1	IIT Library, Delhi	x	x	√	x	x
2	IIT Library, Kharagpur	x	x	√	x	x
3	IIT Library, Madras	x	x	√	x	x
4	IIT Library, Roorkee	x	√	x	x	x
5	IIT Library, Kanpur	x	x	√	x	x
6	IIT Library, Bombay	x	x	√	x	x
7	IIT Library, Indore	x	x	√	x	x
8	IIT Library, Jodhpur	x	x	√	x	x
9	IIT Library, Ropar	x	√	x	x	x
10	IIT Library, Mandi	x	√	x	x	x
11	IIT Library, Hyderabad	x	x	√	x	x
12	IIT Library, Gandhinagar	x	x	√	x	x
	Total	0	3	9	0	0

- There is an urgent need to develop dynamic library websites which will compatible with web 2.0 technologies include web forms in each library 2.0 service.

- The library should run short term courses, organize training or instructional programs on Web 2.0 tools for the students, staff and faculty members.

- Web 2.0 tools, such as blogs, RSS, social networking sites and Wikis, should be used with well-defined purposes and standard norms. This step will improve the reliability of Web 2.0 tools and improve the participation of patrons in activities of libraries.

- Instruction should be given by the libraries for users about web 2.0 tools used in libraries on their homepages of Library.

- University Grant Commission (UGC) of India with the help of INFLIBNET formulate the national Web 2.0 standards to adopt web 2.0 tools and meet the new challenges of next generations library 2.0 and set up Web 2.0 content steering committees to strengthen their controls on information contents.

Table 9. Ranking of 20 IIT's, Libraries in India based on checkpoints used

Sr. No.	**Name of IIT's, Libraries in India**	Total Web 2.0 Tools used	Rank No.
1	IIT Library, Bombay	12	1
2	IIT Library, Hyderabad	10	2
3	IIT Library, Kharagpur	07	3
4	IIT Library, Madras	07	3
5	IIT Library, Delhi	06	4
6	IIT Library, Roorkee	06	4
7	IIT Library, GandhiNagar	06	4
8	IIT Library, Kanpur	05	5
9	IISER Library, Indore	05	5
10	IIT Library, Mandi	05	5
11	IIT Library, Jodhpur	02	6
12	IIT Library, Ropar	02	6

18. CONCLUSION

In conclusion, it can be said that Web 2.0 tools can and has led a powerful paradigm shift in the dissemination and deliverance of education clearly. With the proliferation and adoption of Web 2.0 tools by IIT libraries in India, it is becomes important to help familiarize library users with these tools. Learning about such tools prepare library users for better communications and effective use of library resources and future employment avenues. However, use of these tools library may face challenges related to availability, accessibility and acceptance of technology and difficulty in enthusing and motivating library users to use and leverage such tools.

It is concluded from the study that the IIT's Libraries in India are well aware with the modern concept like Web 2.0 technology and its various tools. IIT's libraries in India always are using such type of various web 2.0 tools for to provide library services to its users. If we see the table number 04, we can find out that Social Networking like Facebook, Gmail, Rediffmail, MySpace, Blog, twitter, LinkedIn etc. are mostly used in the IIT's Libraries in India. As well as Instant messaging, YouTube, Video, Podcast, RSS feeds also becoming very popular day to day among the libraries and library users.

So it is the urgent need today that Librarians should try to adopt new web 2.0 technology and arrange proper training for library professionals so as they become able to provide better library services to the users.

19. FURTHER RESEARCH STUDY:

More research should be undertaken on the following aspect of Web 2.0 tools and Libraries.

- To find out how these tools are being used by the academic and public libraries.

❖ Further studies need to be undertaken regarding the users' feedback of web 2.0 tools and Library 2.0 services by the academic libraries.

REFERENCES

1. Abram, Stephen.(2005). Web 2.0-Huh? Library 2.0, Librarian 2.0. Information Outlook 9.12 .Accessed on 26.11.2011, http://www.sla.org.
2. Anderson, Paul (2007). What is Web 2.0? Ideas, technologies and implications for education. JISC Technology and Standards Watch, Feb 2007." .Accessed on 26.11.2011 http://www. jisc.ac.in.uk/media/document.
3. Dasgupta, Diptima., Dasgupta, Rudranil (2012). Structure and behavior of social networks Linked to Web 2.0. CSI Communications 36(1),5.
4. Ibrahim, Rabiu & Oxley, Alan. Proposed Development Methodology for Higher Education and Library Mash-ups. Accessed on 28.1.2011 http://www.ieeexplore.org.
5. Kumar, Haris, Mishra B.K. & Kulshrestha, S.B.(2010) .Using Blogs for Extending Library services. University News 48.12, 16-31pp.
6. Lihitkar, Shalini (2012). Application of Web 2.0 for Enhancement of Library and Information Services (Focus on Weblog). Presented in UGC refresher course in Library and Information Science at Department of LISc, RTM University, Nagpur 2012.
7. Lihitkar, Shalini (2012). Information Systems and Networks in India. New Delhi: Today and Tomorrows Printers and Publishers,
8. Lihitkar, Shalini (2010). Establishing a Virtual Reference Services. DESIDOC Journal of Library and Information Technology 31(1),31-34pp.

9. Madhusudhan, M.& Nagabhushanam,V (2012). Web-based Library services in University Libraries in India: an analysis of librarians' perspective. The Electronics Library 30(5).569- 588.Accessed on 26.11.2011http://www. emeraldinsight. com.

10. Ramesh Babu & others (2010). Recent developments in Web 2.0: New Avenues for LIS Professionals. Information Age. 4(3) (2010). 05

11. Sinha, Ravi (2012). Meet the Challengers: With social networking skills becoming almost a requisite for success these days, can you really afford to put all your eggs in one basket?. Digit Magazine, (March 2012), 88-89.

12. Tim O'Reilly. What Is Web 2.0? Design Patterns and Business Models for the Next Generation of Software. Accessed on 25.09.2012, http://www.oreillynet. com/pub/a/ oreilly/tim/news/.

13. Tripathi, Manorama (2009). "Use of Web 2.0 tools by academic Libraries: A Reconnaissance of the International Landscape." Accessed on 01.3.2012,www.iatul.org/doclibrary /public/conf-proceeding/2009/tripathi-text/pdf.

Chapter 10

Availability of ICT infrastructure in the University Libraries of West Bengal, India

Arup Kumar Mondal, Dr. Amit Kumar Bandyopadhyay

[1]Scholar (Pt. time), DLIS, BU and Librarian Bolpur Sailabala Girls High School, Bolpur, Birbhum West Bengal, India

[2]Dept. of Library & Inf. Sc. University of Burdwan West Bengal, India

ABSTRACT

The present study is an attempt to asses the way in which the University libraries in West Bengal have responded to changing information environment and developed their infrastructure toward use of ICT. The findings show that ICT infrastructure in the university libraries of West Bengal is still at different stages of development. Most of the libraries have minimum infrastructure for the implementation of ICT, but they face the problem of manpower.

KEYWORDS

ICT, Value–added information service, Video-conferencing, Intranet, Infrastructure, Human resource.

1. INTRODUCTION

Ever dynamic ICT has changed the traditional process of library activities and this situation stresses the necessity for the university libraries to be part of the ICT based information world. ICT offers ample opportunities for libraries to automate the traditional activities, implement efficient and effective library cooperation and resource sharing networks, develop institutional repositories, provide value-added information services and initiate capacity building programmes for library staff and library users.

But implementing ICT in University libraries necessitates a varied set of technological tools. ICT encompasses a wide range of technologies such as telephony, cable, satellite, TV and radio, video-conferencing, digital technologies, computers, Internet, World Wide Web, & intranet and software application.

Indian Statistical Institution (ISI) library for the very first time started computerisation in West Bengal. Later, some special libraries of West Bengal started computerisation. But computerisation began in the State University libraries of West Bengal at the beginning of 90's. Due to lack of proper ICT infrastructure facilities computerisation of the University libraries started late. Now the situation has totally changed. Most of the University libraries in West Bengal have minimum ICT infrastructure and ready for ICT application. The present study is an attempt to asses the way in which the University libraries in West Bengal have responded to changing information environment and developed infrastructure to use ICT.

2. AIMS AND OBJECTIVES OF THE STUDY

The aim of the study is to investigate the availability of information and communication technology infrastructure in the general University libraries in West Bengal.

3. SCOPE & LIMITATION OF THE STUDY

The Universities of West Bengal established after 2000 (2008 onward) are yet to develop proper basic infrastructure. Due to this reason and also considering limited time and money, the state funded general university libraries of West Bengal established before 2000 are included in the study. There are seven such State Funded General Universities in West Bengal as shown in Table 1 (Sl.no.1-7).

4. REVIEW OF RELATED LITERATURE

Kannappanavar and Vijayakumar1 (2001) have made a survey on the use of hardware and software facilities in University of Agricultural science libraries in Karnataka. The results reveal that none of the University libraries at the time of study were having databases and full implementation of IT applications in their libraries. Though the agricultural university libraries are having hardware and software facilities to some extent, the results are not reaching the clientele. It recommends that the librarians should approach the university authorities to train the library personnel on IT application and approach funding agencies like INFLIBNET and ICAR for their library automation and provide IT based information services to their clientele. Suku and Pillai2 (2005) made a survey to assess the status of automation in the university libraries of Kerala. A structured questionnaire was used to elicit data from the Librarian/Librarian in-charge of the Central libraries of six universities. The survey mainly cover various aspects of library automation such as information technology infrastructure, in-house activities, information services and their usage, manpower development, and budget. The study also deals with the role of INFLIBNET Centre in supporting the automation activities of university libraries. It is seen that library automation has been rather slow in Kerala due to various reasons like absence of University Librarian in most of the libraries and lack of adequate qualified professional staff. 50% of university libraries in Kerala introduced comprehensive automation of housekeeping activities. LAN facility is available in all university libraries. All university libraries in Kerala are using computers for their services. All the libraries, without any exception, are using only personal computers for the entire range of automation activities. The survey also reveals that all university libraries have conducted sufficient number of training programs to its staff members before acquiring the new technology. Walmiki & Ramkrishnegowda3 (2009) have surveyed on the status of ICT infrastructure in six selected university libraries of Karnataka and revealed that, ICT infrastructure in the university is still at different stages of development. Insufficient computer hardware and software and lack of internet facilities with required bandwidth indicate that university libraries of Karnataka are yet to establish extensive facilities required for efficient information access. Hence, there is an urgent need on the part of the university libraries in Karnataka to plan implement and develop ICT infrastructure to be fit in facing the challenges ahead of them. Dhanavandan, Esmail & Nagarajan4 (2011) have surveyed on Information Communication Technology (ICT) Infrastructure Facilities in Self-Financing Engineering College Libraries in Tamil Nadu. From the result of the findings of the study it could be noted that 87.14 % of the libraries have client and work station facility. Among them, the IBM, HP and Intel dominate. Most of the libraries do not have CD ROM servers in their libraries but majority of the libraries have CD Tower facility in their respective environment. 95 per cent of the colleges use the library application software in their libraries. More than half of the libraries use digital library software.

5. METHODOLOGY

Survey method has been used to conduct the study. To complete the survey structured questionnaire was prepared and e-mailed to the librarians/in-charges individually in the month of September 2012. In-spite of several telephonic communications little response is received. Later the university libraries were visited physically for data collection.

6. DATA ANALYSIS METHOD

Collected data has been analysed with the use of simple statistical techniques. Data has been presented in the form of tables and/or charts using MS-Excel 2007 software.

Table 1. State Funded General Universities Affiliated to UGC in W.B.

Serial No.	Name of University	Year of establishment	Location
1	Universiyu of Calcutta(C.U.)	1857	Kolkata
2	Jadavpur University(J.U.)	1955	Kolkata
3	University of Burdwan(B.U.)	1960	Burdwan
4	University of Kalyani(K.U.)	1960	Kalyani
5	Rabindra Bharati University(R.B.U.)	1962	Kolkata
6	Universiyy of North Bengal(N.B.U.)	1962	Siliguri
7	Vidyasagar University(V.U.)	1981	Midnapore
8	University of Gour Banga(G.B.U.)	2008	English Bazar
9	West Bengal State University(W.B.S.U.)	2008	Barasat
10	Aliah University(A.U.)	2008	Kolkata
11	Sidho Kanho Birsha University(S.K.B.U.)	2010	Purulia & Bankura
12	Presidency University(P.U.)	2010	Kolkata

Table 1 shows State Funded General Universities Affiliated to UGC in W.B. It is seen from the table that, there are only 12 state funded general universities having in West Bengal. Out of 12 , the questionnaires were sent to 7 university libraries as the last 5(from serial no. 8-12) universities viz. G.B.U, W.B.S.U., A.U., S.K.B.U. & P.U. are established very recently and are just in the process of developing infrastructural facilities, academic programmes and libraries.

7. DATA ANALYSIS AND INTERPRETATION

7.1 Library Budget

Table 2 shows budget provision for the development of ICT infrastructure in three consecutive years. It is seen from the table that out of seven state funded general university only two universities have kept provision to spend amount on the improvement of ICT infrastructure in the library. It is also seen that BU have no budget provision on ICT in 2011- 12

Table 2. Library Budget (in Rs. lakhs)

Name of University	Information Technology		
	2009-10	2010-11	2011-12
C.U.	-	-	-
J.U.	-	-	-
B.U.	1.00	1.00	-
K.U.	-	-	-
R.B.U.	-	-	-
N.B.U.	-	-	-
V.U.	20.00	20.00	20.00

7.2 Availability of Computers

Table 3 depicts availability of computer facility in the state funded general university libraries of West Bengal. Out of seven university libraries C.U. & V.U. have same no. of 12 server machine followed by J.U. with 7 server machine N.B.U. with 4 server machine. On the other hand C.U. has largest no. of 4000 client computers followed by J.U.

having 150 & N.B.U. having 95 client computers. Out of total no of 4436 computers there are only 42 (0.95%) server machines and remaining 4394 (99.05%) client computers. At a glance the result shows that C.U., J.U., N.B.U. & V.U. are in a very good position in the sense of availability of computers.

Table 3. Availability of Computers in the Library

Name of universities	Type of computer		Total
	Server machines	Client computers	
C.U.	12	4000	4012
J.U.	07	150	157
B.U.	03	37	40
K.U.	01	21	22
R.B.U.	03	35	38
N.B.U.	04	95	99
V.U.	12	56	68
Total	42	4394	4436
Percentage	0.95%	99.05%	100%

7.3 Hardware Facilities

The data gathered from questionnaire responses regarding the Infrastructural Facilities available in the university libraries under study is presented in Table 4. It is seen from the Table that all the 7(100%) libraries have Printer, Scanner, UPS, Xerox (Photocopier). 5 (71%) Libraries have Barcode reader & CD/DVD player, 4 (57%) libraries have LCD Projector. 3(43%) libraries have Digital camera, Fax, Overhead Projector & T.V. Only J.U. library has RFID Technology.

In addition to the items mentioned in the table, J.U. library has separate Preservation unit, V.U. has unified storage CCTV system & K.U. has Barcode printer as well as Lamination Machine. Since most of the libraries have the basic Infrastructural Facilities, they would facilitate the introduction & use of IT in the libraries.

Table 4. Accessory Hardware Facilities in the Libraries

Name of University	Digital camera	Printer	Fax	Scanner	Barcode reader	RFID technology	Projector		UPS	T.V.
							Overhead	LCD		
C.U.	Y	Y	N	Y	Y	N	Y	Y	Y	Y
J.U.	Y	Y	N	Y	N	Y	Y	Y	Y	N
B.U.	N	Y	Y	Y	Y	N	N	Y	Y	Y
K.U.	N	Y	Y	Y	Y	N	N	N	Y	N
R.B.U.	N	Y	N	Y	N	N	Y	N	Y	Y
N.B.U.	Y	Y	Y	Y	Y	N	N	Y	Y	N
V.U.	N	Y	N	Y	Y	N	N	N	Y	N
Total	3	7	3	7	5	1	3	4	7	3
Percentage	43%	100%	43%	100%	71%	14%	43%	57%	100%	43%

Y=Yes; N=No;

7.4 Software used

Kinds of software used by the libraries are depicted in Table 5. Regarding Operating system, as seen from the table that all the 7 (100%) university libraries used Windows followed by Linux with 6(86%) libraries and Dos with 3(43%) libraries. About 6 (86%) libraries used Soul as library management software. CDS/ISIS & Libsys are used by only 1(14%) library as library management software. On the other hand 5 (72%) libraries used open source software Despace as digital library management software. C.U. used customised software as it despatch the digitisation work to a private agency. 6(86%) libraries used MS-SQL as a backend DBMS. Only one library used POSTGRESQL as DBMS. There is a variation in using antivirus software. Out of 7 university Libraries, antivirus software NORTON & KASPERSKY are used by 2(29%) libraries. Other libraries used single different kinds of software such as RBU used QUICKHEAL; J.U. used EKILL & V.U. used TRENDMICRO.

It is seen from Table 6 that, all the libraries have WAN & Internet facility. Six (86%) libraries have LAN & 4(57%) libraries have WIFI connection. In addition all the libraries are linked to either one or two National as well as International library network. Statistics also shows that all the libraries (100%) are linked with UGC's INFLIBNET. J.U. &

N.B.U. are linked with both INFLIBNET as well as DELNET. K.U. is the only University which is linked with ERNET & INFLIBNET.

Table 5. Software used by the library

Name Of University	Operating System			Library Management			Digital Library				DBMS		Antivirus				
	DOS	WINDOWS	LINUX	CDS/ISIS	SOUL	LIBSYS	GSDL	DSPACE	FEDORA	CUSTOMISED	MSSQL	POSTGRE SQL	NORTON	KASPERSKY	QUICKHEAL	EKILL	TRENDMICRO
C.U.	N	Y	Y	N	Y	N	N	N	N	Y	Y	N	Y	N	N	N	N
J.U.	Y	Y	Y	N	N	Y	N	Y	N	N	Y	N	N	N	N	Y	N
B.U.	N	Y	Y	N	Y	N	Y	Y	N	N	Y	N	N	Y	N	N	N
K.U.	N	Y	N	N	Y	N	N	Y	N	N	Y	N	N	Y	N	N	N
R.B.U.	Y	Y	Y	Y	Y	N	N	N	N	N	Y	N	N	N	Y	N	N
N.B.U.	Y	Y	Y	N	Y	N	N	Y	Y	N	Y	N	Y	N	N	N	N
V.U.	N	Y	Y	N	Y	N	N	Y	N	N	N	Y	N	N	N	N	Y
Total	3	7	6	1	6	1	1	5	1	1	6	1	2	2	1	1	1
Percentage	43	100	86	14	86	14	14	72	14	14	86	14	29	29	14	14	14

Y=Yes; N=No;

7.5 Campus Wide Data Communication Network

Table 7 demonstrate the number of Terminals provided to the user from the library. It is seen from the table that there are 433 terminals available for use by the users of 7 state funded general university libraries of west Bengal. Out of 7 responded libraries C.U. has the largest number of 150(35%) terminals followed by J.U. with 100(23%) terminals and N.B.U. with 95(22%) terminals. K.U. has a very small no. of 2(0.5%) terminals available for their users.

Table 6. Availability of Campus Wide Data Communication Network

Name of Universities	WIFI	LAN	WAN	INTERNET	Link with regional/national/international network		
					INFLIBNET	DELNET	ERNET
C.U.	Y	Y	Y	Y	Y	N	N
J.U.	Y	Y	Y	Y	Y	Y	N
B.U.	N	Y	Y	Y	Y	N	N
K.U.	N	N	Y	Y	Y	N	Y
R.B.U.	Y	Y	Y	Y	Y	N	N
N.B.U.	N	Y	Y	Y	Y	Y	N
V.U.	Y	Y	Y	Y	Y	N	N
Total	4	6	7	7	7	2	1
Percentage	57	86	100	100	100	29	14

7.6 Terminals Provided To Users

Table 7. Number of Terminals Provided to the Users from the Library

Name of Universities	Number of terminals	Percentage
C.U.	150	35
J.U.	100	23
B.U.	34	8
K.U.	02	0.5
R.B.U.	28	6
N.B.U.	95	22
V.U.	24	5.5
Total	433	100

7.7 Internet Connectivity

The table 8 shows the details of internet connection regarding the type of connection, service provider & bandwidth. It is seen from the table that most of the libraries have leased line connection from different internet service provider (ISP). B.U. & J.U. Libraries have VSAT connection. Except K.U., all other universities have connected through the ISP Tata communications (previously VSNL). Out of 7 universities 4 universities have been connected through more than one Internet

Service Provider. Like C.U. & B.U. have connected with ERNET & Tata Communication, R.B.U. with ERNET & NICNET and N.B.U. with Reliance & Tata communication. The Table also shows that except J.U. all other universities are having 1 Gbps bandwidth facilities.

Table 8. Internet Connectivity through ISP & its Bandwidth

Name of University	ERNET			NICNET			RELIANCE			TATA COMMUNICATION		
	* L. L.	VSAT	BANDWIDTH	* L. L.	VSAT	BANDWIDTH	*L .L.	VSAT	BANDWIDTH	* L. L.	VSAT	BANDWIDTH
C.U.	Y	N	1 Mbps	N	N	-	N	N	-	Y	N	1 Gbps
J.U.	N	N	-	N	N	-	N	N	-	Y	Y	70 Mbps
B.U.	Y	Y	4 Mbps	N	N	-	N	N	-	Y	N	1 Gbps
K.U.	N	N	-	N	N	-	N	N	-	Y	N	1 Gbps
R.B.U.	Y	N	2 Mbps	Y	N	1 Gbps	N	N	-	N	N	-
N.B.U.	N	N	-	N	N	-	Y	N	16 Mbps	Y	N	1 Gbps
V.U.	N	N	-	N	N	-	N	N	-	Y	N	1 Gbps

- L.L.=LEASED LINE

8. FINDINGS

1. Only NBU provides open access library facility to its users. All the other libraries provide either closed / mixed access facility to its users.

2. Most of the University libraries under study have given less importance on implementation of ICT.

3. C.U., J.U., N.B.U. & V.U. are in a very good position in the sense of availability of computers.

4. Since most of the libraries have the basic Infrastructural Facilities, they would facilitate the introduction & use of IT in the libraries.

5. All the libraries have WAN & Internet facility. Six (86%) libraries have LAN and 4 (57%) libraries have WIFI connection. In addition all the libraries are linked to either one or two National as well as International library network.

9. CONCLUSION

ICT infrastructure in the university libraries of West Bengal is still at different stages of development. Most of the libraries have minimum infrastructure for the implementation of ICT, but they face the problem of manpower. Another problem that has come out from the interrogations with the library staff members is that existing manpower are not competent enough with ICT. Hence there is an urgent need to plan properly about the required manpower for providing ICT based library service.

REFERENCES

1. Kannappanavar, B. & Vijayakumar, M., 2001. Use of IT in University of agricultural science libraries of Karnataka: A Comparative study. Desidoc bulletin of information technology, 21(1), pp. 21-26.

2. Suku, J. & Pillai, M., 2005. Perspectives on automation of university libraries in Kerala. Journal of Academic Librarianship, 31(2), pp. 151-159.

3. Walmiki, R. H. & Ramakrishnegowda, K. C., 2009. ICT infrastructure in University libraries of Karnataka. Annals of Library and information studies, Dec., Volume 56, pp. 236- 241.

4. Dhanavandan, S. S., Esmail, M. & Nagarajan, M., 2011. Information Communication Technology (ICT) Infrastructure Facilities in Self-Financing Engineering College Libraries in Tamil Nadu. Library Philosophy and Practice (e-journal).

5. http:// www.wbsche.ac.in/

Chapter 11

Knowledge Sharing through Social Networking Sites (SNS) among Undergraduate Students in a College Library, affiliated to Anna University, India

Chandran Velmurugan[1], Natarajan Radhakrishnan[2]

[1]Department of Library and Information Science, Periyar University, Salem- 636 011, Tamilnadu, India

[2]Department of Library and Information Science, Periyar University, Salem- 636 011, Tamilnadu, India

ABSTRACT

This paper aims to create awareness about social networking sites (SNS) and how it is used to share knowledge among the user community and also deals with to promote effective access to online information resources. A well-equipped and well managed library is the foundation of contemporary education structure. It is said that education with one library services is like a body without soul, a vehicle without engine and building with bricks but no cement. Hence a Library and Information

Research Centers (IRC) are the principal instruments for gathering knowledge. A well-structured questionnaire was employed to test the variables using statistical tools.

KEYWORDS

Social media, social networks, social networking websites, social networking sites, and online communities, College Library, Anna University, India

1. INTRODUCTION

Generally, the terms such as Social media, social networks, social networking websites, social networking sites, and online communities are synonyms and have changed the way we communicate both in institutions as well as individuals. It provides the opportunity to all of us to connect in ongoing conversations with our friends, family, fans, staff and students, parents, alumni, colleagues about which is most essential to them. In recent days, more and more people are moving towards the Internet to keep in touch with friends, family and colleagues. Social networking sites such as Facebook, Flickr, twitter, YouTube, MySpace, Pinterest, Google+ and many more sites make staying involved easier, quicker and even more fun than ever before. Facebook and MySpace are recently the most popular SNS in North America and more to choose. These sites have millions of users with more signing up day by day.

In the knowledge era, technology has had a significant impact on the delivery of library enquiry and reference services, from the introduction of first generation technologies such as email and chat to second generation technologies such as Facebook and Twitter (Benn and McLoughlin, 2013). Barsky and Purdon (2006) defined that Social network sites offer a free and easy way to create personal web pages and fill them with content such as blogs, digital photo graphs, favorite music, short video clips, and much more and also social networks are formed as members link their web pages to those of their friends and search through the vast number of sites in search of new friends who might share common interests. Social networking is used to publicize library events such as gaming nights; to alert users to additions to collections; to provide links to articles, videos, or web content that might prove relevant or helpful to patrons; and to provide a conduit for community information (American Library Association report, 2012). Wikis, blogs, chat rooms, instant messengers, message boards and social bookmarking are technology applications that have been used to facilitate members" interaction, and thus, have been referred to as social networking tools (Jones & Conceicao, 2008). However, sites such as YouTube and Flickr have been identified as primarily for sharing videos and images, rather than social networking (Hoffman, 2009).

1.2 Profile of College and Central Library

Sri Ramanujar Engineering College is one of the leading institutions in Chennai, Tamilnadu, India. It is affiliated with Anna University, Chennai, and accredited by AICTE. The vision of the college is to be a leader in providing quality education in the fields of Engineering, Technology and Management, at International Quality, blended with social and moral values, for the betterment of mankind, through empowerment of youth with special emphasis on rural youth and the mission is to prepare students to become competent professionals, by providing state of the art learning environment with highly motivated faculty Nurture talent and innovation in students for holistic development. The college is running with following departments such as Civil Engineering, Computer Science & Engg, Electronics and Communication Engineering, Electrical & Electronics Engineering, Information Technology, Mechanical Engineering, Science & Humanities and one PG Programme that is Management Studies (MBA).

The library is well stocked with a wide range of resources covering all areas of engineering and related subjects. The institute also subscribes to a number of National and International periodicals and journals; these are excellent sources for updating information. Library also boasts of a Digital Library section where high end computers are connected through high speed internet connection to IIT Library, Springer and IEEE etc.

2. RELATED WORK

A review of related literature is an account of what have been already published on a related topic or subject by eminent scholars, researches and information scientists. Based on the authors" experience, this paper briefly reviews literature to exhibit the significance of social networking sites in libraries.

Gupta and Seth (2015) have carried out the study on awareness of Social Networking Services among the Faculty Members of Degree Colleges of Jammu Division J&K India and the study highlighted the problems encountered by the participants and suggested some remedial measures for improving awareness in SRELS Journal of Information Management, LeBlanc and Kay (2014) conducted study on web 2.0 and social media and applications for Academic Libraries during 2014 to implement in Amelia V. Gallucci-Cirio Library, Fitchburg State University. Researchers found that more or less a wide range of web 2.0 tools have been executed in higher education Libraries and Information Centers around the world and also found communication media such as wikis, blog, multimedia, social networking sites have being effectively used in educational institutions. Chu and Du (2013) examined on social networking tools in academic libraries using online questionnaire based survey with 140 university Libraries among Asia, North America and Europe. The results indicated that most of the library professionals had positive attitude on the usefulness of social networking tools and only few library staff members and user community had perceived positive attitude. The other related studies are some of them for the present study given below.

Jill Benn and Dawn McLoughlin (2013) expressed the study by facing our future: Social Media Takeover, Coexistence or Resistance? In the Integration of Social Media and Reference Services, Chu and Chen (2012) have discussed the SNS Tools which would be useful for libraries based on their study entitled, Social Networking (SN) Tools For Libraries, Seminar at The University of British Columbia Library in the University of Hong Kong, Boroughs (2010) explained in his theses regarding the social networking websites and voter turnout which is available in ProQuest Dissertations and Theses database. Taylor-smith and Lindner (2009) observed using social networking tools to promote e- participation initiatives presented at the Technology, colleges and Community Worldwide online conference in Honolulu, Hendrix et al., (2009) reported their survey on the use of Facebook in academic health sciences libraries in Journal of the Medical Library Association, Jones and Conceicao (2008) in their article social networking tools faster informal learning presented at the 24th Annual conference on distance teaching and learning in Madison. Boyd and Ellison (2007) described the social networking sites and its definition, history and scholarship in Journal of Computer – Mediated Communication. Read (2006) noticed in his papers to the chagrin of its clientele, Facebook expands its membership in the Chronicle of Higher Education, Barsky & Purdon (2006) have carried out the study on introducing Web 2.0: social networking and bookmarking for health librarians in the Journal of Canadian Health Libraries Association.

3. OBJECTIVES OF THE STUDY

The study was conducted to find out the awareness and access pattern of social networking sites (SNS) which are available in the library by the student community of Sri Ramanujan Engineering College. The study was conducted to achieve the following objectives.

- To assess the awareness of SNS among the user community;
- To observe the purpose of using SNS;
- To find out the quantum of time spent in hours and years for using SNS;
- To explore the frequency of using SNS;
- To identify the media sources for using SNS;
- To find out the problems faced by the users while using SNS in Libraries and
- To analyze the level of satisfaction using SNS

4. SCOPE AND LIMITATION

The present study is limited and extent to undergraduate students of Sri Ramanujan Engineering College, affiliated to Anna University, Chennai city to explore the awareness of social networking sites (SNS).

5. METHODOLOGY

To gain an insight into the social networking sites (SNS) usage by student community, data was collected from students of central library at Sri Ramanujan Engineering College. The questionnaire contained both open ended and close-ended questions. Out of 300 students, two hundred and eighty nine questionnaires were returned dully filled in by the user"s community and the overall response rate was 96.34 per cent. The gathered data were classified, analyzed and tabulated by using simple statistical methods. This study covers only undergraduate students" community of the college and the study is to know their usage of social networking sites (SNS) for sharing update knowledge in the digital era.

5.1 Analysis

To achieve the goal, the study was conducted at Sri Ramanujan Engineering College central library campus in Chennai. A total number of 300 Students were randomly selected and distributed to students in the college, out 300 questionnaires 289 questionnaires were returned duly filled in which is 96.34 %.

- ***Survey Respondents***

Table 1 Figure 1 shows that the respondent wise distribution of response sheet to fill the questionnaire for analysis purpose for the present study. A well designed questionnaire were employed and distributed to 300 students of six under graduate departments. Out of 300, the maximum number of 50 out of 50 students (16.67%) filled their questionnaire from Electronics and Communication Department, followed by 49 (16.33%) from Civil Engineering, 48 (16.00%) from EEE and Mechanical Engineering, 47 (15.67%) from Computer Science and IT Departments respectively.

- ***Gender wise distribution***

Respondents were grouped into two categories such as male and female to retrieve the data for the present study. It can be observed from the above Table 2 Figure 2 indicates that the majority of 158 (54.67%) respondents are from male category and the remaining 131 (45.33%) are from female category. It is found that most of the male respondents are very much interested to show their performance in filling the questionnaire to share their knowledge.

Table 1. Survey Respondents

S. No	Department	Total Students =300		Survey Respondents =289	
		No of Students	Total %	No of Respondents	Total %
1	Civil Engineering	50	16.67	49	16.33
2	Computer Science & Engg	50	16.67	47	15.67
3	Electronics and Communication Engineering	50	16.67	50	16.67
4	Electrical & Electronics Engineering	50	16.67	48	16.00
5	Information Technology	50	16.67	47	15.67
6	Mechanical Engineering	50	16.67	48	16.00
Total		300	100.0	289	96.34

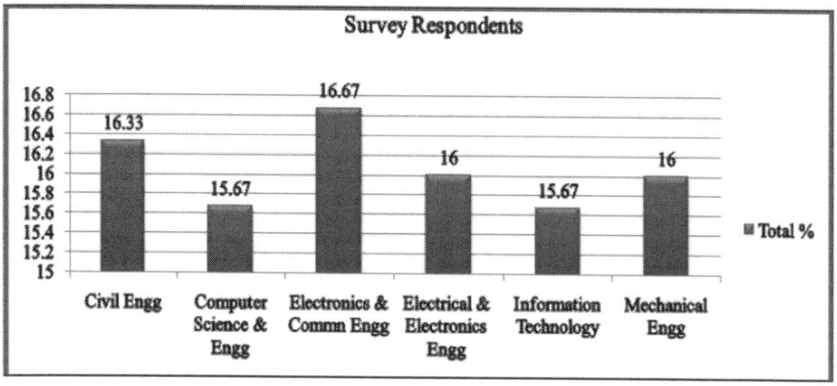

Figure 1. Survey Respondents

Table 2. Gender wise distribution

S. No	Sex	Response count n=289	Response frequency %
1	Male	158	54.67
2	Female	131	45.33
Total		289	100.0

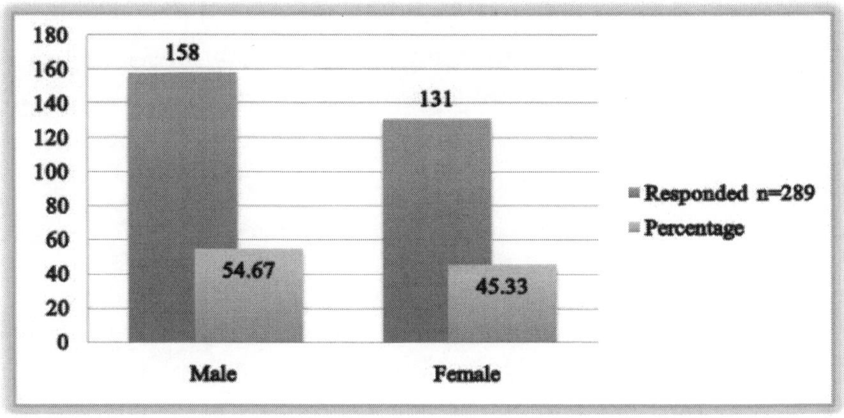

Figure 2. Gender wise distribution

- ***Awareness of SNS***

Respondents were asked to know about the awareness of social networking sites (SNS) and the results shows that the majority of 286 (98.96%) user community are well aware about the SNS and the very least number of 3 (1.04%) users are not aware of it and do not know how to use effectively and how to become a member of SNS.

- ***Frequency of using the SNS***

Respondents were asked to know about the frequency of using the social networking sites (SNS) during the study period. Out of 289 users, the highest number of 132 (45.68%) students to visit social networking sites daily and followed by 88 (30.45%) users twice in a week and the lowest number of 6 (2.07%) users to visit the SNS rarely.

- ***Time spent (in hours) surfing SNS***

It is observed from the above Table 5 and Figure 5 represents that the time spent in terms of hours using social networking sites by the users. Out of 289 respondents, the maximum number of 187 (64.71%)

respondents are spent time surfing SNS 30 minutes per day, followed by 42 (14.53%) respondents spent time surfing SNS 30 to 45 minutes, 33 (11.42%) respondents are spent time surfing SNS 45 to 60 minutes and the least number of 9 (3.12%) respondents spent time surfing SNS only one year as they are not much awareness about SNS and its services.

- ***Time spent (in Years) surfing SNS***

It is inferred from the above Table and Figure indicates that the length of time spent in years using social networking sites by the users. Out of 289 respondents, the majority of 121 (41.86%) respondents are well aware of SNS and spent time surfing SNS three to four years, followed by 82 (28.37%) respondents spent time surfing SNS two to three years and the least number of 9 (3.12%) respondents spent time surfing SNS only one year as they are not much awareness about SNS and its services.

Table 3. Awareness of SNS

S. No	Opinion	Response count n=289	Response %	Frequency
1	Yes	286	98.96	
2	No	3	1.04	
Total		289	100.0	

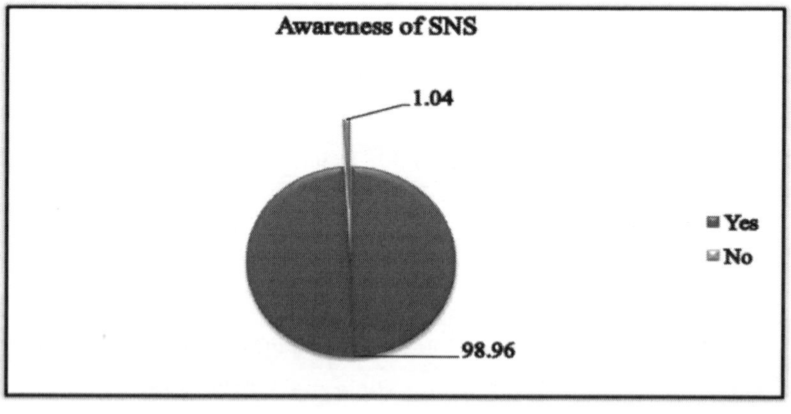

Figure 3. Awareness of SNS

Table 4. Frequency of using the SNS

S. No	Frequency	Response count n=289	Response Frequency %
1	Daily	132	45.68
2	Weekly	46	15.92
3	Twice in a week	88	30.45
4	Monthly	17	5.88
5	Rarely	6	2.07
Total		289	100.0

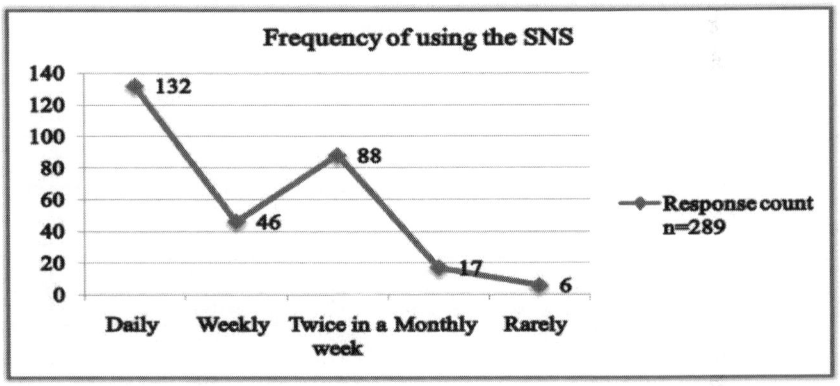

Figure 4. Frequency of using the SNS

Table 5. Time spent (in hours) surfing SNS

S. No	Time Spent Hours (per day)	Response count n=289	Response frequency %
1	0-30 minutes	187	64.71
2	30-45 minutes	42	14.53
3	45-60 minutes	33	11.42
4	One to Two hrs.	18	6.22
5	More than Two hrs.	9	3.12
Total		289	100.0

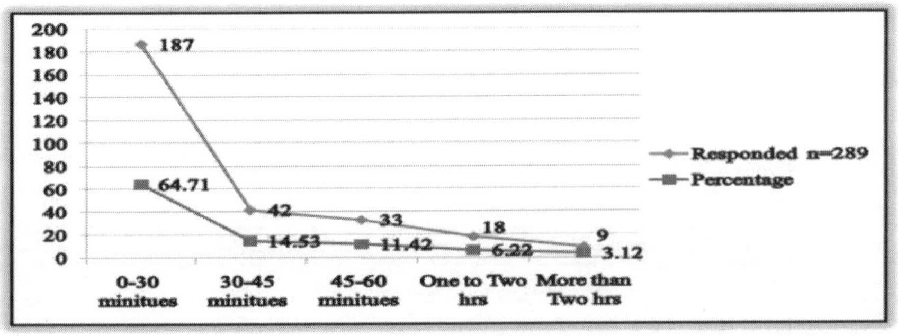

Figure 5. Time spent (in hours) surfing SNS

Table 6. Time spent (in Years) surfing SNS

S. No	SNS Use in Years	Response count n=289	Response frequency %
1	One year	9	3.12
2	Two years	64	22.15
3	Two to Three years	82	28.37
4	Three to Four years	121	41.86
5	More than Four years	13	4.50
Total		289	100.0

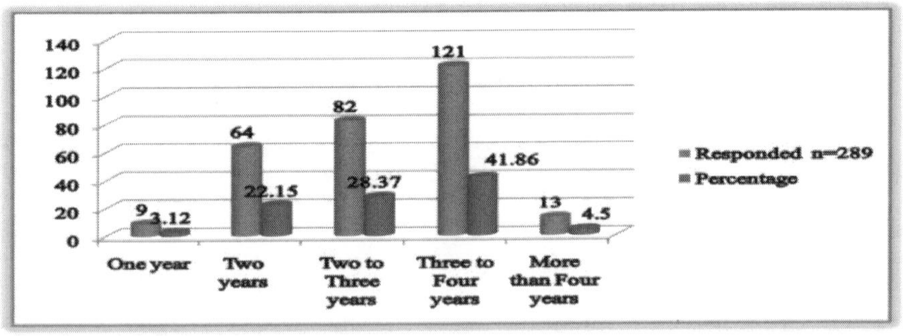

Figure 6. Time spent (in Years) surfing SNS

- ***Media Source for surfing SNS***

Media is very essential tool to share and update information among the researchers and scientists. In this present study, the respondents asked to note about the media sources (such as computer laboratory, university library, and mobile laptop and cyber café) where they can surfing social networking sites to share their knowledge among them. It could be observed from the table 7 (fig.7) represents that out of 289, most of the respondents are preferred to share their views and ideas by using mobile phones (43.60%) which placed first rank and followed by laptop (33.91%), cyber café (15.58%), only the least number of respondents are preferred to share their knowledge and informant by using University Library (2.76%).

Table 7. Media Source for surfing SNS

S. No	Media source	Response count n=289	Response frequency %
1	Dept. Computer Lab	12	4.15
2	University Library	8	2.76
3	Mobile	126	43.60
4	Laptop	98	33.91
5	Cyber café	45	15.58
Total		289	100.0

- ***Purpose of using SNS***

Table 8 (fig.8) indicates that the purpose of accessing social networking sites (SNS) by the user community during the period of study. The results reveal that the maximum number of 92 (31.38%) respondents are browsing SNS for sharing their information to their friends and relatives and followed by 75 (25.95%) respondents are using social networking sites for making new friends to have good relationship in terms of academically and socially. The next 45 (15.58%) user community by means of SNS for academic purpose and followed by 36 (12.45%) respondents are using SNS for their employment oriented purpose and the least number of 14 (4.84%) user community use the SNS site for the purpose of mere entertainment.

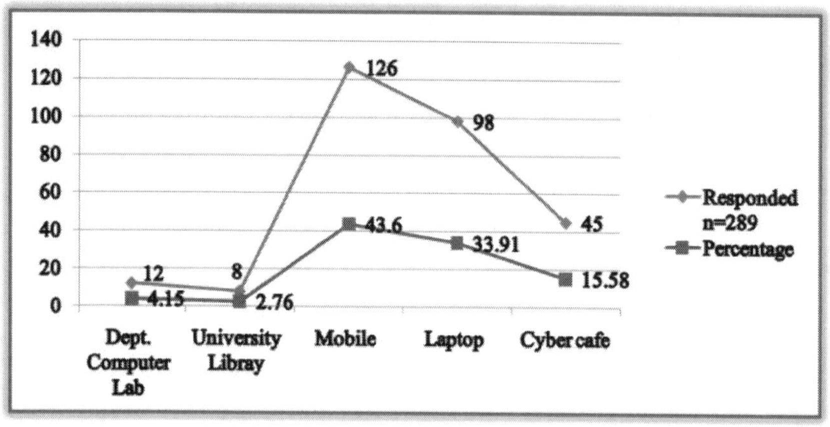

Figure 7. Media Source for surfing SNS

Table 8. Purpose of using SNS

S. No	Purpose	Response count n=289	Response frequency %
1	Sharing Information	92	31.83
2	Academic Purpose	45	15.58
3	For Chatting	27	9.35
4	Making new friends	75	25.95
5	For Entertainment	14	4.84
6	For Job Information	36	12.45
Total		289	100.0

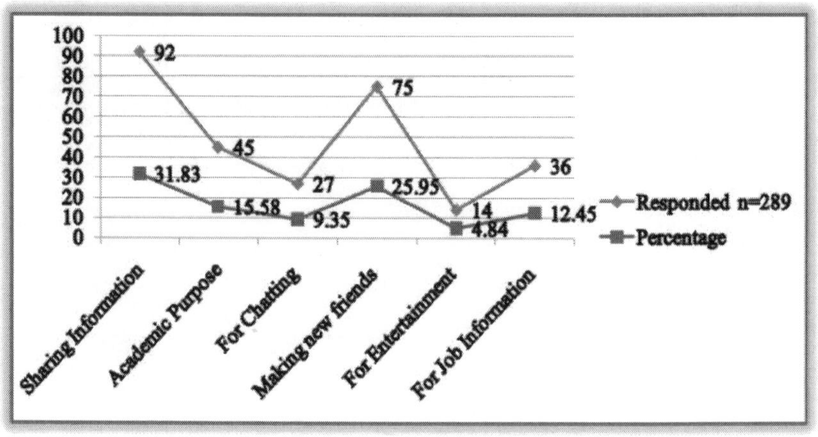

Figure 8. Purpose of using SNS

- ***Advantages of using SNS***

It is inferred from the Table 9 (fig.9) represents that the advantages of using social networking sites for student community. Generally, a number of advantages as well as disadvantages are using the SNS. In this regard, we have taken into consideration only advantages of SNS for the present study. Hence, as per the above table, the majority of 99 (34.26%) user communities have taking the advantage for making new friends academically and intellectually to share their information and followed by 62 (21.45%) respondents stated that they take an advantage in lieu of communication skills to learn more knowledge other than the subjects and followed by 55 (19.04%) users taking as advantage of SNS to communicate with elite people who guide them in an innovative way and more over to get suggestions and ideas in terms of research work, assignments and project oriented. Only few numbers of 12 (4.15%) respondents are using for the purpose of SNS to update current affairs apart from subjects.

Table 9. Advantages of using SNS

S. No	Advantages	Response count n=289	Response frequency %
1	For Communication skills	62	21.45
2	For Writing skills	24	8.30
3	For General knowledge	37	12.80
4	To update current affairs	12	4.15
5	To make new friends academically	99	34.26
6	To communicate with elite people who guide us	55	19.04
Total		289	100.0

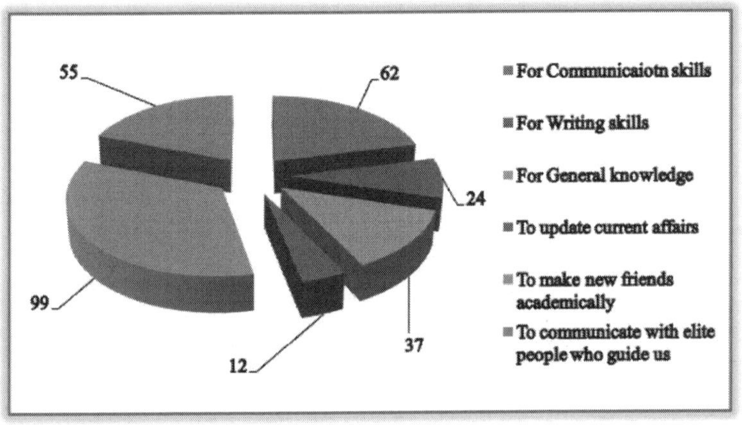

Figure 9. Advantages of using SNS

- ***Type of information uploaded in SNS***

Table 10. Type of information uploaded in SNS

S. No	Uploaded Information	Response count n=289	Response frequency %
1	Related with Subject	41	14.18
2	Updated current news	83	28.72
3	Related General knowledge	109	37.72
4	Related useful video /image	37	12.81
5	Related useful Fun activity	19	6.57
Total		289	100.0

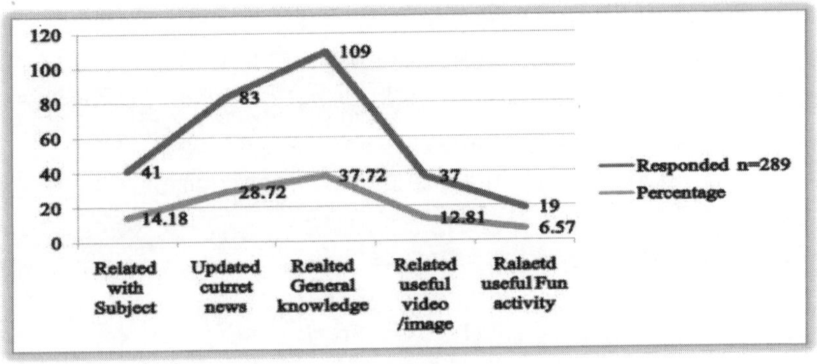

Figure 10. Type of information uploaded in SNS

- ***Importance of SNS for career employment***

Table 11. Importance of SNS for career employment

S. No	Importance of SNS	Response count n=289	Response frequency %
1	Yes	249	86.16
2	No	26	8.99
3	No Idea	14	4.85
Total		289	100.0

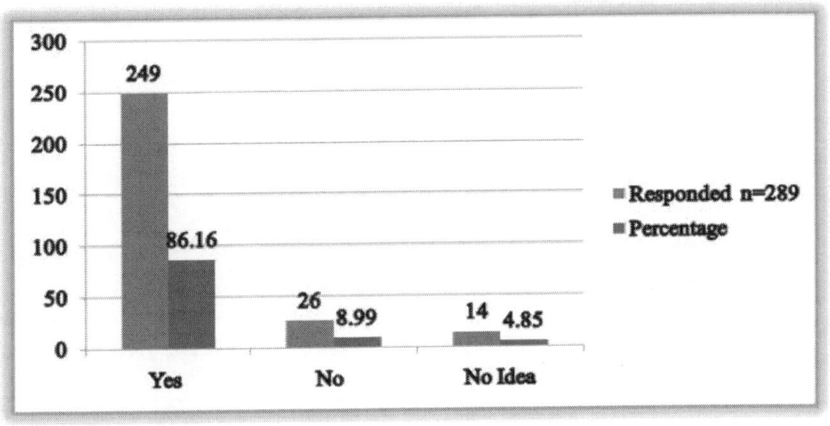

Figure 11. Importance of SNS for career employment

It is noticed from the above table 10 (fig.10) that most of the respondents 249 (86.16%) are provided their opinion that the SNS sites are greater impact in terms of job information whereas the least number of 26 (8.99%) respondents are viewed that social networking sites are very least importance regarding the carrier choice.

- ***Problems while using SNS in Libraries***

Table 12. Problems while using SNS in Libraries

S. No	Problems while using SNS	Response count n=289	Response frequency %
1	Poor Internet connectivity	102	35.30
2	Lack of timing	38	13.15
3	Lack of Trained Library staff	86	29.75
4	Lack of maintenance	35	12.11
5	Financial support by Organization	28	9.69
Total		289	100.0

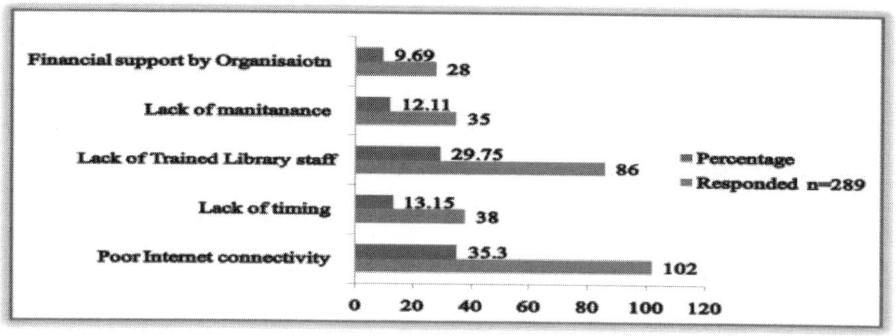

Figure 12. Problems while using SNS in Libraries

It is inferred in the table 12 (fig.12) represents that the problems are facing by the respondents while using SNS in libraries and information centers. The uppermost numbers of 102 (35.30 %) respondents are facing poor Internet connectivity while using in SNS in libraries and followed by 86 (29.75%) respondents are meeting problems with lack of trained library professionals while asking any doubts regarding the SNS sites from them. The next 38 (13.15%) are having lack of time to use SNS and only 28 (9.69%) facing financial support by organization.

- ***Level of Satisfaction using SNS***

S. No	Satisfaction Level	Response count n=289	Response frequency %
1	Highly Satisfied	92	31.84
2	Satisfied	82	28.37
3	Partially satisfied	71	24.57
4	Dis satisfied	26	8.99
5	No comments	18	6.23
Total		289	100.0

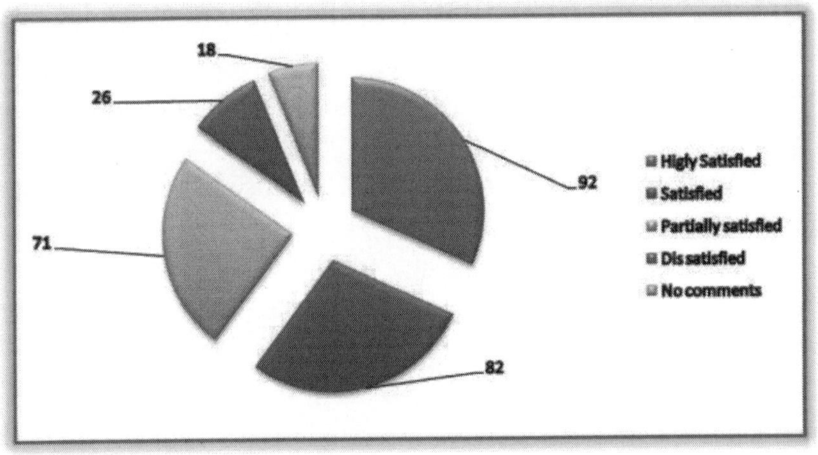

Level of Satisfaction using SNS

It can be observed from the above table that the level of satisfaction using the Social networking sites by the user community during the period of study. It shows that the highest numbers of 92 (31.84%) users are highly satisfied by means of SNS services and followed by 82 (28.37%) respondents are satisfied. The next 71 (24.57%) respondents indicated that they are partially satisfied with SNS Services and 26

(8.99%) student communities are dissatisfied with SNS sites and their advantages and the small numbers of 18 (6.23%) respondents are stated that there are no comments in terms of social networking sites.

6. FINDINGS AND CONCLUSION

We can conclude that Social networking sites (SNS) are one of the emerging thrust areas in the electronic age. Based on the analysis, the results reveal that almost 99 percent of user community are well aware about the SNS and only one percent users are not aware of it and do not know how to use effectively. Out of 289 respondents, the majority i.e. 41.86% of respondents are well aware of SNS and spent time surfing SNS three to four years and the least number i.e. 3.12% of respondents spent time surfing SNS only one year as they are not much awareness about SNS and its services.

The majority of 34.26% user communities have taking the advantage for making new friends academically and intellectually to share their information and only few numbers (4.15%) are using for the purpose of SNS to update current affairs apart from subjects. The greatest numbers of 109 (37.72%) respondents are uploading information in SNS related with general knowledge while very less 19 (6.57%) respondents are uploaded information related with useful fun activity.

The huge number of 102 (35.30 %) respondents are facing poor Internet connectivity while using in SNS in libraries and the least number of 38 (13.15%) are having lack of time to use SNS and only 28 (9.69%) facing financial support by organization. The highest numbers of 92 (31.84%) users are highly satisfied by means of SNS services and followed by 82 (28.37%) respondents are satisfied whereas the small numbers of 18 (6.23%) respondents are stated that there are no comments in terms of social networking sites.

7. RECOMMENDATIONS

We know that information has the different parameters such as verbal communication, print communication, electronic way of communication and recently digital age. We have seen that from Stone Age to Digitized Age, those who have knowledge would pass it to the next generation, the receiver would adopt the knowledge, analyze it, improve it, add new changes to it and then pass it on to the next generation and the cycle thus goes on (www.preservarticles.com). It is found that based on the findings of the research, few of the recommendations are as follows:

For sharing library resources and maintain the library culture, social networking services will support with existing information services.

Shared and combined tools such as social networking sites like LinkedIn, Face book, Twitter, blogs like blogger, web blog, Forums like discussion forums, library link etc. would be used to initiate project works, quiz contests, brainteasers, puzzles for user community effectively.

REFERENCES

1. American Library Association (2012). The 2012 State of America's Libraries. Available from World Wide Web: http://www.ala.org/news/mediapresscenter/americaslibraries/soal2012, accessed on 6th June 2015.

2. Barsky & Purdon (2006). Introducing Web 2.0: social networking and bookmarking for health librarians. Journal of Canadian Health Libraries Association, 27, 65-67.

3. Boroughs, B. (2010). Social networking websites and voter turnout. (Master thesis). Available from ProQuest Dissertations and Theses database. (UMI No. 1475359)

4. Boyd, D.M. & Ellison, N.B. (2007). Social network sites: definition, history, and scholarship. Journal of Computer-Mediated Communication, 13(1), 210-230.

5. Chu, Samuel Kai-Wah. & Du, Helen. (2013). Social Networking Tools for Academic Libraries. Journal of Librarianship & Information Science, 45(1), 64-75.

6. Chu, Samuel Kai-Wah and Chen, Dora Yu-Ting. (2012), Social Networking (SN) Tools For Libraries, Seminar at The University of British Columbia Library, The University of Hong Kong.

7. Hendrix, D., Chiarella, D., Hasman, L., Murphy, S., Zafron, M.L. (2009). Use of Facebook in academic health sciences libraries. Journal of the Medical Library Association, 97(1), 44.

8. Information, retrived from: www.preservarticles.com, accessed on August 2014.

9. Jill Benn and Dawn McLoughlin. (2013). Facing Our Future: Social Media Takeover, Coexistence or Resistance? The Integration of Social Media and Reference Services, available at: http://creativecommons.org/licenses/by/3.0/

10. Jones, B., & Conceicao, S. C. O. (2008, August). Can social networking tools foster informal learning? Paper presented at the 24th Annual conference on distance teaching and learning, Madison, WI.

11. Khan, Shakeel Ahmad, and Bhatti, Rubina (2012). Application of social media in marketing of library and information services: A case study from Pakistan. Webology, 9(1), Article 93. Available at http://www.webology.org/2012/v9n1/a93.html

12. LeBlanc, Linda and Kim, Kay (Kwahng) (2014), "Web 2.0 and Social Media: Applications for Academic Libraries." Information Security and Computer Fraud, vol. 2, no. 2 Page: 28- 32. (http://pubs.sciepub.com/iscf/2/2/2/) accessed on 14/04/2015.

13. Read, B. (2006). To the chagrin of its clientele, Facebook expands its membership. The Chronicle of Higher Education. 53(5), A.35.

14. Sangita Gupta and Anuradha Seth (2015) awareness of Social Networking Services among the Faculty Members of Degree Colleges of Jammu Division J&K India, SRELS Journal of Information Management, 52(2), 2015, 125–130.

15. Taylor-Smith, E. & Lindner, R. (2009, April). Using social networking tools to promote eparticipation initiatives. Paper presented at the Technology, Colleges and Community Worldwide Online Conference, Honolulu, HI

Index

N

O

R

S

T

U

V

W

Y